Routledge Revivals

Rural Change and Planning

Originally published in 1996 *Rural Change and Planning* describes the turbulent changes that have occurred in rural England and Wales since the outbreak of the First World War. The book describes the changes from an agriculturally-dominated countryside to one which has had to increasingly adapt to urban pressures. Looking at the changes chronologically, the book provides an integrated history of rural planning in the twentieth century and the developments which have taken place within the State, which has facilitated those changes. The book looks at the social and economic impacts of two world wars on agricultural communities, and the pressures of industry, new settlements and the effects of recreation on rural landscapes.

Rural Change and Planning

England and Wales in the Twentieth Century

Gordon E. Cherry and Alan Rogers

Routledge
Taylor & Francis Group

First published in 1996
by E & FN Spon

This edition first published in 2018 by Routledge
2 Park Square, Milton Park, Abingdon, Oxon, OX14 4RN
and by Routledge
711 Third Avenue, New York, NY 10017

Routledge is an imprint of the Taylor & Francis Group, an informa business

A Library of Congress record exists under LCCN: 94202257

ISBN 13: 978-1-138-34420-4 (hbk)
ISBN 13: 978-0-429-43833-2 (ebk)

RURAL CHANGE AND PLANNING

RURAL CHANGE AND PLANNING

England and Wales in the Twentieth Century

GORDON E CHERRY and ALAN ROGERS

E & FN SPON
An Imprint of Chapman & Hall
London · Glasgow · Weinheim · New York · Tokyo · Melbourne · Madras

Published by E & FN Spon, an imprint of Chapman & Hall, 2–6 Boundary Row, London SE1 8HN, UK

Chapman & Hall, 2–6 Boundary Row, London SE1 8HN, UK

Blackie Academic & Professional, Wester Cleddens Road, Bishopbriggs, Glasgow G64 2NZ, UK

Chapman & Hall GmbH, Pappelallee 3, 69469 Weinheim, Germany

Chapman & Hall USA, 115 Fifth Avenue, New York, NY 10003, USA

Chapman & Hall Japan, ITP-Japan, Kyowa Building, 3F, 2-2-1 Hirakawacho, Chiyoda-ku, Tokyo 102, Japan

Chapman & Hall Australia, 102 Dodds Street, South Melbourne, Victoria 3205, Australia

Chapman & Hall India, R. Seshadri, 32 Second Main Road, CIT East, Madras 600 035, India

First edition 1996

This book was commissioned and edited by Alexandrine Press, Oxford

Typeset in 10/12½ pt Times by Cambrian Typesetters, Frimley, Surrey

Printed in Great Britain by St Edmundsbury Press, Bury St Edmunds, Suffolk

ISBN 0 419 18000 1

A catalogue record for this book is available from the British Library

♾ Printed on permanent acid-free text paper, manufactured in accordance with ANSI/NISO Z39.48-1992 and ANSI/NISO Z39.48-1984 (Permanence of Paper).

To the memory of

Gerald Wibberley

1915–1993

colleague, mentor and friend

This book, sadly, must serve as Gordon Cherry's penultimate piece of scholarship in the field of planning history, to which he made so many distinguished contributions over more than a quarter of a century – not least, by his joint editorship of the present series and of the periodical *Planning Perspectives*. His contributions to this book convey so well the meticulous scholarship and his acute sense of social and political changes which impacted on the developments of planning policies. His sudden and untimely departure leaves a huge gap which it will be difficult, even impossible, ever to fill.

Peter Hall
The Bartlett
University College London

CONTENTS

PREFACE

Towards the close of a century during which rural areas have experienced profound changes in landscape, social and economic circumstances and in living conditions, there remains a deep and abiding interest in the past and a lively concern for the future. *Rural Change and Planning* is about this transformation of the countryside in England and Wales between 1914 and the present day: it recounts what has changed, it explains how, it suggests why, and reveals the extent to which the State itself has been an agent of change.

For our readership we have students in mind from a variety of disciplinary backgrounds: typically, geography, town planning and environmental studies, suitably informed from economics, politics, sociology, history and ecology. With more than 80 years to cover, the appeal to history is self-evident, but there is a wider point behind the historical emphasis, particularly for geographers and planners. We fear that both disciplines have grown a-historic in their outlook. Planners have imbued a recent educational culture that shows little sustained interest in the past; it discounts the importance of recognizing that contemporary problems have origins and that policy solutions have consequences. Geographers are not immune from these criticisms; the teaching of contemporary geography is pointedly lacking in historical perspective. Neither planning nor post-modernist geography stand aloof in the sands of time.

We present a multiple-focus sketch of the countryside. There is already a voluminous rural literature, but it is largely uni-focused, typically around issues such as agriculture, recreation, landscape and environment. It is timely to present rural questions in the round: it is neither possible nor desirable to fragment the countryside agenda into self-contained, constituent parts. Of course any account, descriptive or analytic, has to be structured in its presentation, and this book is no exception, with its specific chapters. But we are at pains to present a synoptic story which invites cross-referencing, both implicit and explicit. There is a seamless web of relationships which is unprofitable to unstitch.

Geographers and environmentalists have a common interest in landscape and land use. The historical geographer regards the landscape as a palimpsest, literally a record of sequential features still observable or deducible at the present day. The skeleton of the landscape is in its basic rock and soil which gives topography and natural vegetation. But human form and colour have been added over countless centuries. This is where the landscape historian and historical geographer make their contributions. Hoskins (1955) was concerned 'with the ways in which men have cleared the natural woodlands; reclaimed marshland, fen and moor; created fields out of a wilderness; made lanes, roads and footpaths; laid out towns, built villages, hamlets, farmhouses and cottages, created country houses and their parks; dug mines, and made canals and railways; in short, with everything that has altered the natural landscape' (p. 14).

Our present twentieth-century review adds to that understanding of landscape change: the continued clearance of the woodland, and its replacement in selected parts; the on-going reclamation of wetland and moor; new field patterns, new agricultural features, new land uses; the making of new highways; the creation of new towns and the expansion of others; new buildings of all description; new mining operations and the closure of old ones; and the closure of canals, the dismantling of railways, the opening of airports. During the twentieth century a pattern of new landscapes has been formed, adding a further footnote to the historical geography of England that H.C. Darby prepared in 1936 and revised in 1973.

Addressing so many different readership interests, and covering such a lengthy period of time, obliges us to make a presentation which, while descriptive, is yet a critical commentary. Our chosen form is synoptic narrative with the objective of telling a sequential story, where particular themes and periods can be identified. For ease of readership unduly copious referencing and footnotes have been avoided, though major source material is properly identified. The broad canvas demands selectivity in subject matter; wider reading can readily supplement the main lines of the story. Above all, the emphasis is that the course of rural change can only be seen in context and there is constant reference to the wider shifts in social, economic and political affairs.

An introductory chapter reviews a number of cross-cutting themes from the past; it reminds the reader of important contemporary issues; and it provides a background for understanding developments in twentieth-century town and country planning. The historical narrative proper begins with a sketch of the countryside in 1914; this permits an acknowledgement of the legacy of earlier periods of rural history by portraying the essential features at the outbreak of the First World War. The next three chapters unfold chronologically: 1914–18, 1918–39, and 1939–45. The two World Wars broke continuities; they disrupted patterns and threads from the past and introduced new factors, hence each is called an hiatus. The 20 years between are neatly contained; they provided important experiences addressed by the formulators of post-war policy. After 1945 a thematic approach is adopted, the individual chapters being concerned with agriculture, industry, landscape protection and recreation, rural communities and housing, and aspects of land use planning. A final chapter recapitulates on the major themes.

We should like to recognize the services provided by our respective departments – the School of Geography at the University of Birmingham and the Environment Section, Wye College, University of London – and, for their exemplary secretarial services, we should like to thank Beryl Bryant and Valerie Eden. Finally, our thanks to our families who have lived with the production process through good times and bad.

Gordon E. Cherry and Alan Rogers
May 1995

1

TWENTIETH-CENTURY THEMES

The countryside and its past have a particular appeal which from time to time takes on an emotional intensity. In the First World War poets saw pastoral England in almost mystical terms and resoundingly felt that it was worth fighting for. Rural Wales has its passionate adherents too. In our present day huge numbers in amenity organizations flock to the defence of the countryside from development. The emotional concern has been heightened through the abiding impression that the landscape is destined to experience even more disfigurement than has so far occurred. Hoskins (1955) communicated this view uncompromisingly 40 years ago, and his anti-development bias is shared by a broad cross-section of opinion today. He observed that since 1914

> every single change in the English landscape has either uglified it or destroyed its meaning, or both. Of all the changes in the last two generations, only the great reservoirs of water for the industrial cities of the North and Midlands have added anything to the scene that one can contemplate without pain (p. 298).

Neither the obliteration of former landscapes nor the erosion of ways of life has dulled the contemporary appetite for the rural past. How else is it possible to explain the popularity in recent years of literary re-creations of former village life?: Flora Thompson's trilogy *Lark Rise to Candleford* (1945) or Ronald Blythe's *Akenfield* (1969) for example. Equally enduring have been the excursions into landscape history. W.G. Hoskins (1955) in *The Making of the English Landscape* opened the eyes of the lay person, student and scholar alike with his vivid account of landscape evolution from pre-Roman days to the present. Further, R.E. Moreau (1968) showed the interest that could be extracted for the general reader in respect of one village in Oxfordshire, Berrick Salome, from the wealth of historic data extending from the time of a land grant by Aethelred II in 996.

This deep sense of historic association, the persistent attachment to the countryside, and a widespread regret about the passing of the familiar, all provide the backcloth to the chapters which follow. We feature England and Wales from 1914 to the present, (Scotland and Northern Ireland have sufficiently different features to warrant separate treatment) but it is impossible to isolate these 80 years or so from what had gone before. At the outset therefore it is important to acknowledge a long continuity in which there are recurrent threads.

THE CONTEXT OF THE PAST

Over many centuries two main forces of change have operated on the countryside: the economic drive of agrarian capitalism, with its related social order, and the impact of urban values on countryside interests.

The capitalist landlord emerged early in national history, as seen in woodland clearance, wetland drainage, the recovery of waste and heath, and the enclosure movement which secured more pasture for the valuable wool trade. By the eighteenth century an occupational structure of tenant farmers and wage-labourers was in place, firmly cemented by the progression of the parliamentary enclosure movement. From the 1750s to the 1820s six million acres of land were taken into new ownership through nearly 4,000 Enclosure Acts, promoted in the main by powerful landowners. A long process, whereby land holdings were concentrated and given a hierarchical social structure of landlord, tenant and labourer, had unfolded almost to a conclusion. The apogee was reached with eighteenth-century agrarian capitalism and the highly productive years of the 1860s, characterized by 'high farming', a term synonymous with high production achieved by the judicious application of scientific knowledge and equipment (Chambers and Mingay, 1966). Progressively the landless left the farms and became the working class of the growing industrial towns. Social and political developments have brought about significant changes in the twentieth century, notably in tenurial shifts, but in other respects the essential features of capitalist agriculture remain.

Important aspects of the evolution of agrarian capitalism were related to the application of technological advances to agriculture. Drainage schemes required improved pumping techniques; the reclamation of poor soils needed fertilizers, made available through advances in chemistry. The application of science to farming was progressive, affecting stock breeding, crop yields, drainage and soil enrichment. In the twentieth century these advances would be taken to levels never previously contemplated.

Another feature of agrarian capitalism was its exposure to world competition. As described in chapter 2 home agriculture fell on hard times because of the cheapness of imported food and there was only a partial recovery by 1914. The point to make is that there was a continuity from the nineteenth to the twentieth century in terms of the pattern of world competition. Inter-war trading difficulties continued while post-war problems have related particularly to issues of European surpluses.

The second factor, the imposition of urban values on the countryside, has given rise to diverse forms of exploitation. An early example was the piecemeal invasion of the countryside for residential purposes, as seen in the country house with (from the seventeenth century) its landscaped garden, fashioned according to the dictates of the time. The present day urbanite who takes over a country cottage or retires into a village community has followed in the footsteps of the nineteenth-century town merchant buying a country seat.

A much more substantial form of rural land exploitation came with massive suburbanization. The possibilities of mass transport, successively by train, tram, petrol bus and car, the attraction of healthy living away from the smoky, sunless town, and the increasing social attraction of suburban villadom greatly stimulated the popularity of country living. In the second half of the twentieth century personal mobility, afforded by a massive extension of car ownership and a combination of highway improvements and new road building which has greatly facilitated access and penetration to formerly remote areas, has made rural living altogether more possible for former town dwellers. The perceived social problems of the big cities, including various forms of lawlessness, have reinforced the small town and village as desirable places for family raising where an overall quality of life is attainable. Over time, the result has been that urbanites have treated the countryside as theirs, an alternative place to live in and a resource to take over.

The urbanite's discovery of the countryside for recreation purposes dates from the last quarter of the nineteenth century. The bicycle permitted countryside excursions and hiking, and rambling grew in popularity. The railway offered a means of access to locations further afield. In south-east England certain localities were colonized for informal weekend retreats, as further described in chapter 2. The twentieth century saw an explosion in leisure activities with a countryside focus, the car permitting penetration extensively into areas previously considered remote, for an ever-widening range of venues, interests and experiences.

LESSONS FOR THE PRESENT DAY

During the twentieth century the socio-political structures in rural England and Wales have become increasingly complex. The relative powerlessness of the agricultural worker may have persisted, as Newby (1977) suggested in his study of farm workers in East Anglia, and the inherited privilege of landed property may have remained instrumental in social and political power relationships, but the arrival of new power groups cannot be denied.

The strength of the special interest lobbies is now considerable: they are well organized with large nominal memberships, knowledgeable (often beating off the expert), articulate (sweeping in style, but persuasive against the spokesperson of a local authority department, constrained against giving unguarded comments) and in command of media presentations. On the environmental agenda they have been outstandingly influential. The Council for the Protection of Rural England, the Ramblers, Friends of the Earth, Greenpeace and the like are as monolithic and single-minded in purpose as Trades Union barons of old in the field of industrial relations. Moreover they have popular support, at least to the extent that they can be called upon as guardians of amenity, a role which the public sector once claimed.

These changing power structures suggest that a new balance is being struck between town and country. The present-day approach to land planning has not yet fully responded to this point. The Town and Country Planning Act, 1947 might just as well have been called the Town *versus* Country Planning Act: towns and cities were separate from the countryside and good planning would keep them so. But the reality of post-war metropolitan development has served to blur that distinction; spatially and culturally the urban imprint is now much more extensive, so much so that Cloke's (1983) 'index of rurality', which indicated those tracts that could pretend to show truly rural characteristics, were sharply declining in area, and no doubt they have shrunk further since.

Power relationships have other dimensions too. The future of the English and Welsh countryside is now intimately bound up with decisions taken not only in Brussels about 'European' agriculture, but also in board rooms across the developed world about investment in Britain. It is affected by scientific advances in information technology which affect location decisions in foot-loose industry. It is also influenced by cultural movements the world over which affect attitudes and beliefs which impact on local policies regarding conservation and development. The countryside is in change.

Our review is insistent on the place and role of agriculture in the landscape and in the countryside generally. Throughout the twentieth century, though of reduced significance both in employment terms and from the point of view of its contribution to the national economy, agriculture still remains of social and political importance in rural life, perhaps out of all proportion. It also retains its crucial significance as the shaper of the landscape, visually in its arable or pasture characteristics and its field patterns; with the withdrawal or removal of farming operations the effect on the landscape is far-reaching. Agriculture and rural planning are inseparable.

The place of industry in the countryside has been less consistent, but remains important, and may be becoming more so. A healthy rural economy (beyond agriculture) is eminently desirable because of its consequential effects on community livelihoods. An urban take-over of rural industries (town-based engineering and food processing for example) had already begun in the nineteenth century, a loss which was met only half-heartedly with a nostalgia for craft workshops and small-scale plant. For much of the twentieth century there was little change in the disproportionate imbalance of urban and rural industrial location. But rural areas now appear much more attractive to capital than for many years (perhaps indeed since the early days of the Industrial Revolution when many of the early factories had rural locations). Flexible systems of production permit a new look at what rural areas have to offer, and the service sector now readily favours non-urban centres for its operation.

A deep attachment to rural living remains another century-long feature. The sentiment has been a wholesome one of social intimacy and the rejection of anonymity, the suitability of small settlements for family upbringing, an association with health and a 'natural environment' and a gentler pace of life. Alternative perceptions as to remoteness, backwardness, poor facilities and disadvantage appear as of less importance. Depopulation may have been a feature of some particular parts, especially in Wales and the English uplands, but by and large villages have not been deserted; rather they have been re-inhabited by other occasional or permanent residents. Moreover, post-war schemes to abandon settlements which had lost their former economic *raison d'être* were bitterly resisted and overturned. The villages and small towns of England and Wales are prized assets and many people strive to have a share in them.

The declining distinction between urban and rural in spatial and cultural terms has been noted. Yet rurality is still very important and this has been highlighted, particularly in the last quarter of a century by the environmental movement. The extraordinary attachment to an unchanging landscape (no matter sometimes how ordinary, and notwithstanding that it was created and reshaped by recurrent human endeavours over the centuries) is an enduring feature of the twentieth century. It accounts for the emotional intensity of the National Parks campaign 60 years ago, the similar force of the environmental lobby at the present time and the continuing vigour of the demand for right of access – the right to roam over any kind of country (as Marion Shoard (1987) puts it, 'this land is our land'). The more urban England and Wales become, the more their scenic heritage and unchanging landscape qualities are valued.

As a result, another key feature of the century remains the conflict between protectionist and developmental pressures. The countryside has become a battle zone for two contending parties: those who regard rural land as a resource to be exploited, used and developed, and those who see it as a heritage to be preserved and protected. The issue comes to a head in areas of landscape value – scenic, historic, archaeological or ecological – but such is the anti-development point of view, any landscape is now seen as worthy of protection. Development means more roads, cars, pollution and noise; buildings disfigured – a landscape changed is a landscape ruined. The battle lines are drawn between implacable opponents, while those who are concerned with public policy know that compromises have to be struck. The role of planning as a mediator to resolve conflict is a difficult one to discharge.

A unique feature of the twentieth century has been the substantial involvement of the State in countryside affairs. The State was not absent of course from such matters in the nineteenth century; the Corn Laws, the operation of the Poor Law and the Game Laws are testimony enough for that. But the present century throws up an entirely new perspective, and recent change in the countryside is inexplicable without reference to the vicissitudes of protectionism versus free trade in agriculture, and public sector intervention versus the operation of the free market in terms of land. This aspect requires further comment, lengthy enough for a separate section.

THE CONTEXT FOR TOWN AND COUNTRY PLANNING

Over the last 150 years there has been a remarkable extension of State intervention in society at both local and central levels. The broad outline is as follows. During the second half of the nineteenth century the foundations of the modern civil service were laid, local government was progressively reformed and a range of effective *ad hoc* bodies created. Collectivism gradually superseded individualism. The problems posed by an urban industrial society led to the enlargement of State activities: it was necessary to prevent harm to people, hence measures to secure public health and factory regulation were taken by mid-century. It was also argued that unrestrained competition had consequences harmful to the general good of the community: large-scale public undertakings of gas, water and transport proliferated in the big cities by the end of the century. Public facilities were provided by the State: parks, art galleries, libraries, baths and wash houses. Education and policing became major extensions of the State. There was also State involvement in housing standards, and, by the end of Victoria's reign, early arrangements for the provision of housing for the working classes had been introduced.

There was commensurate growth in the machinery and manpower of central government. Boards were created to take over functions performed earlier by the Privy Council and the Home Office: Local Government (1870), Agriculture (1889) and Education (1913). There was an expansion in the number of civil servants. In local government the salaried officials also increased their number and became functionally more important, taking over a wider range of jobs and supplanting the influence and role of the local politician. County Councils were created in 1888, taking over certain functions previously the responsibility of the Quarter Sessions. Consolidation of an array of special bodies into Urban and Rural Districts and Parish Councils from 1894, and the creation of the London County Council in 1889, gave England and Wales a local government map that survived throughout the next century. The efficacy and strength of local government was widely supported, not least by the political parties which found local issues fertile ground in which to exercise power. The dual system of governance (local and central) which had emerged, suited rather than created British conditions: the system was robust and adaptive and it provided conditions whereby political compromise could usually be effected.

With structures of governance in place, there followed the increasing tendency of the State to engage more and more in aspects of national life: to protect, to provide and to regulate. A professional society, with power commanded by experts and specialist knowledge, came to dominate the twentieth century (Perkin, 1989). The election of a Liberal government in 1906 provided a fillip to this trend as the early years of the century saw various aspects of social reform appear on the political agenda. The twentieth-century welfare state took early recognizable form with legislation covering, for example, education, pensions and insurance. The first town planning legislation also dates from this time, in 1909 (Cherry, 1982), as does the setting up of the Development Commission, which marks the beginning of systematic State intervention in economic and social affairs in rural areas.

The impact of the Great War was undoubtedly to increase the areas of government involvement in national life: no war which demanded conscription, direction over the country's economy and control of its food supply and distribution could result in less. After 1918 some State engagement fell away, as with agricultural subsidies, but others were taken up, as with council housing.

But the intellectual thrust behind planning as a State activity in social and economic affairs was gathering pace. The Fabian Society, founded in 1884, worked unremittingly towards a democratic socialist state. Nineteenth-century *laissez faire* Liberalism was discarded in favour of social democratic Liberalism, and the Labour Party shed guild socialism for State socialism. All main political parties were ultimately persuaded by J.M. Keynes' thesis that governments could spend their way out of recession by investment (such as commissioned work on public projects) and by encouraging consumers to spend (by increasing the available money supply).

The notion that society might be comprehensively planned signalled the arrival of the Age of Planning. As an ideology it came to dominate much of the twentieth century, particularly between the 1930s and the late 1970s.

Planning as an article of faith gathered its adherents (Cherry, 1993). There was an a-political appeal in that planning was seen as a modern industrial technique of management where experts were in control, technical efficiency considered much superior to the inefficiency of democratic institutions. Planning promised order as opposed to the chaos that was so much in evidence, the application of Reason offered more than the hand of Chance. Research and conscious choice were preferable to the manipulations of financiers, and surely the waste, the inefficiency and duplication inherent in the market could be bettered by competent, technical expertise? Although assumptions ran ahead of proof, these were the arguments which increasingly held sway. Professional experts would replace the capitalists of industrial society and public authorities would manage the all-embracing activity of planning.

The article of faith had of course its other side. If the very complexity of modern economic organization was a reason why planning was essential, was it not also a reason why it might be impossible? It was argued that the operations in such a system would be too numerous, various and complicated to permit planned control and that the sheer weight of central administration would choke enterprise.

In the event, theoretical speculation was overridden by the exigencies of global armed conflict. With the fall of Chamberlain the Coalition government was formed on 10 May 1940. Advocacy for central planning won the day. Resources were mobilized for total war and within a year a national command economy was established. The experiences of the Second World War proved conclusive; planning won the war (how else would such a massive operation as landing men on the Normandy beaches have been undertaken?), and planning would win the peace. Between 1942 and 1945 the Coalition government committed its successors in a series of White Papers to major initiatives in the fields of education, health, housing, employment, social security, industry and environmental planning. The requirements of post-war reconstruction deemed planning necessary: the rebuilding of blitzed cities, the return of a war economy to peace conditions, assistance to the industries and the regions which had suffered in the depression years, the recovery of a countryside to prosperity, the health, housing and welfare of a country which looked forward to a better world. All this meant central control over the operation of private markets in land, housing and development by agencies of the State. Public authorities would be the benevolent shapers of post-war Britain.

By the end of the war a political consensus promised a planned period of reconstruction. Advisors of the time were confident and certain: Dudley

Stamp (1946), geographer and Director (since 1930) of the Land Utilisation Survey of Great Britain declared that 'What is needed is a comprehensive *national* plan of balanced land use which will accord the necessary *guidance* to the local authorities: guidance which in the past has been sadly lacking' (p. 79). Nothing less than the national planning of land by central government – guiding rather than controlling local initiative and private enterprise – was necessary.

For just a few years the preparation of centrally-directed schemes of national improvement was helped along by a conducive, highly disciplined condition of the British people (Hennessy and Seldon, 1987). Affected by the siege economy on the home front or by military service abroad, the population was accustomed to receiving orders and to strict regulation. Sweeping plans and programmes were prepared by the Attlee government in the image of the war effort, but before the close of the 1940s the reforming zeal had been eroded. However, although the collectivist hour might be over, the collectivist State had been established and major features would endure for 20 or 30 years. The basic infrastructure of a planning system put in place during the later 1940s has remained in place; for the countryside it was applied through town and country planning, agriculture, nature conservancy and National Parks. Underlying assumptions as to urban containment and preservation of the countryside remain to all intents and purposes unchallenged; attempts to dislodge the green belt are firmly repulsed. Nature conservation, despite the twists and turns of institutional arrangements, is strongly supported.

By the early 1970s town and country planning came under fierce scrutiny. Disenchantment set in with the results of the planning system, particularly it must be said in respect of urban areas rather than the countryside. But the allegations of slow, cumbersome procedures and insensitivities on the part of an over-weaning bureaucracy fuelled a general sentiment that well-meaning professionals could, after all, ride roughshod over the interests of others and that experts, not necessarily neutral, could no longer be trusted. Town and country planning seemed to lose its earlier conviction, increasingly buffeted by conflicting interests in the process of consultation and negotiation. The planning system lost vision and purpose as it became systematized and over-regulatory; it was used by the public to stop things happening rather than to achieve. Special interest groups were advantageously empowered in a system which often dealt with negatives and which was costly to operate.

Meanwhile the post-war consensus crumbled. The broad cross-party support for policies based on Beveridge and Keynes was abandoned. Callaghan in the mid-1970s proclaimed the end was nigh; the New Right and the Thatcher governments of the 1980s confirmed it. Individualism and freedom replaced collectivist, egalitarian values.

The frontiers of the State were rolled back through measures of privatization and deregulation as market forces were re-established as the best regulations of economic activity.

Today's change of outlook from 50 years ago is remarkable. We have quoted Stamp's clear, unequivocal advocacy of 1946 proclaiming the virtues and necessity of a planned land use. Compare Donald Denman (1980) a Cambridge land economist writing from the stable of the Centre for Policy Studies:

> There was a hearty arrogance about the recommendations for the use of centralised coercive power. Barlow, sitting nearer in time to the memories of a free market economy, was far more cautious than were Scott and Uthwatt. Scott and his colleagues blandly assumed that the planners would know 'best' what to do with land, better than the owners and farmers would. Uthwatt distinguished the 'proper' planning of the planners from the 'improper' (i.e. uncoordinated) planning of the landowners. (p. 51).

Denman was led to conclude that planning had become process without purpose.

All this constitutes the changing background of twentieth-century planning; the town and country planning system and its related spheres of government activity have variously impacted on and interrelated with the economy and community affairs of the countryside. It forms an inseparable adjunct to the situations described in the following chapters. Once fully established (by virtue of the 1947 Act), the system has rested on some important constants. For example the system remains an obligatory one, with local planning authorities bound by statute to make development plans of one nomenclature or another. It is a democratic system because control over it is in the hands of local councils, popularly elected, and ultimately Parliament. It is an administrative system, planning appeals being referred to the Secretary of State for the Environment, not to courts of law. It is a system rooted in the principle that land ownership *per se* confers no right to develop, refusal of planning permission attracting no right of compensation. And it remains a system in which agriculture and forestry are still substantially excluded from its provisions.

Other constants have revolved around the singular importance accorded to agriculture and arrangements to secure its productive capacity; the resolve to separate as far as possible urban from rural and to enhance the basic qualities of the countryside including village and small town life; and the special protection given to areas of scenic heritage, and environments of scientific importance. Subsequent chapters will examine how far these ambitions have been met. Interestingly, tentative attempts to liberalize rural planning in the 1980s (to match some initiatives in the urban areas taken to encourage market conditions to prevail, such as with Enterprise Zones, Simplified Planning Zones and areas covered by Urban Development

Corporations) have largely failed. There has been strenuous resistance to attempts to relax planning constraints over agricultural land, the green belt is more vigorously defended than ever, housing land release is tightly regulated, and even small-scale development through farm tourism is regarded suspiciously. Free market orthodoxy is conveniently overturned when other interests are at stake.

2

THE COUNTRYSIDE IN 1914

In order to set the scene for what follows, this chapter begins with some pointers as to the main features of the countryside at the onset of the Great War. We consider its appearance and what it was like to live in, farm, work in and visit, and the extent to which social and political attitudes to rural issues were in flux. To provide this historical context, we sketch the fundamental elements of the historical geography of rural England and Wales at around the turn of the century. This need only be a brief resumé; more detailed reviews are available as in Newby (1987), Howkins (1991) and Cherry and Sheail (forthcoming), amplified by a wealth of other introductory material on the Victorian countryside, notably in Mingay (1981, 1989) and on particular facets of rural life by specialist writers.

The socio-economic base of the countryside had undergone profound transformations throughout the nineteenth century, and this process was to continue. Hence no one year can represent a definitive snapshot in what constituted a complex and evolving pattern of relationships. But the main features can be indicated, and the main ingredients of change identified. Against this background, we can begin to highlight the main topics which will recur throughout the book, thus providing the agenda for twentieth-century countryside change and the course of public sector intervention in social, economic and environmental matters.

A NOTE ON EVIDENCE

By the beginning of the twentieth century data and records were amply available to facilitate the task of land use and landscape reconstruction by later generations of historical geographers. Improvements in statistical compilation and in mapping during the second half of the nineteenth century had already contributed to significant developments in the collection of agricultural statistics, population data, and in cartography.

The compilation of numbers of livestock and the parcels of land for different crops from every county in Great Britain had begun on a regular

The rural scene before the Great War. A view of the village of Holybourne, near Alton, Hampshire, showing the London Road and White Hart Inn, *c.* 1890. (By courtesy of the Rural History Centre, University of Reading)

basis in 1866, earlier opposition to the practice having finally been overcome (Best and Coppock, 1962). The annual returns (responsibility for which was transferred from the Board of Trade in 1889 to the newly constituted Board – later Ministry – of Agriculture, though collection and distribution of the schedules was undertaken by the Board of Customs and Excise until 1919) soon developed greatly in complexity. The twenty-five items for which returns were sought in 1866 had doubled by 1919. Returns for the number of horses were first made in 1869; in the same year flax was added to the list of crops, sugar beet in 1873 and small fruit in 1887; the acreage under orchards was first returned in 1871, for market gardens in 1872 and for mountain and heathland in 1891. In 1884 information was obtained for the first time about certain crop yields.

There is of course the question of the accuracy of the returns. Participation was voluntary and not obligatory until during the Great War, but by the early 1900s non-cooperation was unusual. There must have been omissions, errors and misinterpretations: the distinction between permanent grass and rough grazing caused difficulties, also that between permanent and temporary grass, while market gardens and 'bare fallow' proved hard to define. But whatever their limitations, the returns provide us with a fairly accurate picture of agricultural land use at the beginning of our period.

Population censuses had been held at regular ten-yearly intervals after 1801 (Lawton, 1978). The earlier censuses of 1801–31 were largely single population enumerations, but from 1841, the first census undertaken under the (new) office of the Registrar General, a larger range of information was provided, including age and sex structure, birthplace, occupations and housing. The 1911 census made some important additions to the enumeration, with questions on the duration of marriage and the numbers of children born to each marriage, on industry as well as occupation, and on the number of rooms per dwelling.

Throughout the nineteenth century there was a difficulty in that the registration districts (the units of census organization) did not readily distinguish between urban and rural in tabulations. In particular, there was a problem with the classification of suburban populations. The creation of the Urban and Rural Sanitary Districts in 1872 helped somewhat, but the artificiality remained even with the separation of urban and county areas in the local government reorganization of 1889.

Population data helped to assess levels and directions of migration – important at a time of pronounced urban-rural and inter-regional shifts. But the censuses had their limitations in this respect. Seasonal migrants were excluded, for example. The county was the smallest geographical unit for migration data, but such areas were often very diverse in their internal characteristics. Furthermore, migrants were only counted once, even if a number of moves had been made within a decade. A migrant was not counted at all, should a return be made to an original birthplace within that period.

Source material for the historical geographer also came from maps and their changing records of natural and man-made features. The origins of the Ordnance Survey lay in the later years of the eighteenth century, but the modern authority of the Ordnance Survey to survey and provide maps dates from the Ordnance Survey Act, 1841 (Harley, 1975; Seymour, 1980). In 1855 the Board of Ordnance was abolished, its powers transferred to the War Office; in 1870 it was transferred to the Board of Works, and from there to the Board of Agriculture twenty years later. During the second half of the century the whole country was surveyed at 1:2500 scale, the landscape of Victorian Britain progressively recorded in a series of maps of high quality.

Rural land use was suggested by representation of vegetation types. The earliest county maps were remarkably detailed, with orchards, gardens, woods, fir plantations, rough pasture, marshes and a variety of vegetative cover all distinguished by separate symbols. But by the 1880s and 1890s there was a clear trend towards simplifying and omitting minor vegetation features. However, other landscape elements remained, with the landscape faithfully recorded: chalk pits, clay pits, quarries and spoil heaps, archaeological features and railways, and field, parish and administrative boundaries.

Data and mapping were supplemented by an important novelty of the nineteenth century, the photograph, widely in evidence after the late 1850s. Helped by the introduction of the 'dry-plate' process in the later years of the century, the camera began to supplant the sketch book in capturing the visual scenes of the countryside. There was soon a commercial use and the photographic record of Francis Frith in 'Frith's Postcards' was virtually nationwide – a compilation we can return to today with instruction.

HISTORICAL GEOGRAPHY: THE ESSENTIAL FEATURES

Population

The population of England and Wales stood at 36.1 million in 1911, having risen from 17.9 million in 1851; against this increase, however, the number living in Rural Districts fell from 8.9 million to 7.9 million over the same period. As we have already observed, administrative boundaries might not have distinguished accurately between rural and urban, and in fact it has been estimated that the 'true' urban population may have been almost 700,000 higher in 1851 and 2.3 million higher in 1911 (Law, 1967). The scale of under-representation in the shift of population from rural to urban during this time was therefore perhaps quite pronounced. Most rural communities in England experienced a peak in their population around mid-century; by 1911 massive depopulation had reduced population levels often to below their early nineteenth-century figures (Lawton, 1983). As to the nature of this migration, Ravenstein (1885) had already suggested the predominance of short-distance movement, with only occasional transfers over long distances. By the end of the century migration patterns were dominated by inter- and intra-urban movements.

Rural-urban migration reflected a drift from the land, particularly noticeable during the years of agrarian depression. Widespread concern at depopulation and its long-term effects was a contributory factory to the setting up of the Development Commission in 1909. The number of agricultural workers in England and Wales stood at 643,000 in 1911; sixty years earlier it had been 965,000. A push-pull factor was at work, hardship in farming paralleled by urban attractions including improved working conditions, better pay and easier access to housing. The selective nature of the migrant population was evident: the youngest, most able of the rural work force were migrating not just to other parts of this country, but also to territories overseas. Young women migrated too, drawn into domestic service.

The loss of population from the rural areas had been neither uniform across England and Wales nor continuous in respect of particular areas (Coppock, 1973). But of the sustained flight from the land there could be no doubt: in 1911 less than one person in eleven was engaged on the land, whereas in 1851 the ratio had been more than one person in five. Sometimes particular areas experienced a phenomenal rate of rural-urban exchange; one example is that of South Wales in the last quarter of the nineteenth century and into the next. While rural Wales lost its natural increase (in some parts population numbers actually declined), urban Glamorgan boomed. In 1911 its population of 1.25 millions was greater than that for the whole of Wales in 1851.

Overall, the boundaries of rural England and Wales had been progressively rolled back by urban growth. The situation in 1914 was the culmination of change over half a century and more. The nineteenth century had been one of remarkable urban growth, the impact of which fell heavily and directly on the countryside, with a significant transfer of land from rural to urban. A hierarchy of country towns changed in 'pecking-order' as many settlements stagnated, while others expanded, some notably with new industries, particularly associated with the railway. Some townships had expanded in rural areas, often virtually from nothing, as in the case of the railway towns (Crewe, Swindon, Eastleigh and the like) or new manufacturing towns such as Middlesbrough, where the mining of Cleveland ores served to transform a few houses in a salt marsh into a rapidly growing industrial centre. Coastal resorts were growing too: between 1881 and 1911 their aggregate population in England and Wales grew by more than 600,000 to 1.6 million, with the English south coast resorts holding pride of place in terms of their expansion, though in Wales Penarth in the south and Colwyn Bay in the north were beginning to take off.

But the biggest change could be seen in suburban growth around London and the larger cities, new housing areas eating into the countryside as armies of shop workers, clerks, commercial travellers and the rising professional classes sought out fashionable residential districts, newly built on the periphery, often in association with tram or rail networks, and pleasantly associated with woods and fields which were never very far away. The urban fringe spread into the surrounding countryside; in 1914, as now, one important problem area, particularly, in rural England was ever-spreading suburbia (Thompson, 1982). Birmingham had its privileged Edgbaston, Leeds expanded into Chapel Allerton and Headingley, Sheffield's well-to-do sought out the villages of Dore and Totley and built on the high ground between the valleys, Liverpool spread into places like West Derby, Bristol favoured Clifton, Cardiff had its Roath Park, Newcastle its Jesmond and Gosforth and Hull its Cottingham. Around

London where more than two million people lived in its suburban, half-urban half-rural, indeterminate zone, surrounding villages became dormitory communities; here the marks of a rural past, in terms of footpaths and field boundaries, were slow to disappear.

Agriculture

Farming, particularly in England, had experienced prosperity in the 1850s and 1860s. After the repeal of the Corn Laws in 1846 agriculture, responding to a free market, benefited from scientific innovation with regard to fertilizers and breeding, the introduction of mechanization (including the portable steam threshing engine), innovation in farm implements, and widespread drainage projects. The advance of capitalist agriculture produced a Golden Age that transformed a largely self-sufficing domestic industry into a profit-earning manufactory of bread, beef and mutton for a rapidly expanding urban population, better fed than ever before. Of striking significance was the increased acreage of cereal cultivation.

But the tide of agricultural prosperity ebbed in two recessions: 1875–84 and 1891–99. The last quarter of the nineteenth century became a traumatic period which saw 'the transference of agriculture, farming and the landed interest, from a position in the forefront of British polity, economy and society to a relatively minor role' (Perry, 1974, p. 14). Agriculture accounted for 20 per cent of the gross national product in the late 1850s but for only 6 per cent in the late 1890s; it employed over one-fifth of the population in 1851, but less than one-tenth in 1901. R.E. Prothero (later Lord Ernle), writing in 1912, considered the last twenty-six years of Victoria's reign 'a period of agricultural adversity – of falling rents, dwindling profits, contracting areas of arable cultivation, diminishing stock, decreasing expenditure on land improvement' (p. 346). Adverse weather in a succession of unfavourable seasons did not help, but a deeper and more lasting cause for the bad conditions was the changing nature of international trade in food products and a shift in national patterns of food consumption.

Strong overseas competition was first experienced in respect of cereals, with imports of wheat rising markedly as a consequence of the opening up of new wheat-growing areas in Canada, Australia, and Russia, the development of continental railways and improvements in steamship transit. Subsequently, after successful advances in chilling and refrigeration, cheaper meat from the Empire, Argentina and the United States flooded on to the English market. Dairy products particularly from Holland and Denmark were also imported in large quantities.

International transport costs fell sharply and by the beginning of the twentieth century there had been a remarkable transformation in the agricultural scene. Over the period 1909–13 the UK was importing 40 per cent of its meat, 62 per cent of its animal feeding stuffs, 72 per cent of its dairy produce, 73 per cent of its fruit and 79 per cent of its grain and flour for human consumption (Coppock, 1971). Per capita consumption of bread and potatoes fell; that for meat, dairy produce and fruit and vegetables increased.

The overall impact on livestock farming in England and Wales was not too serious; rather, it was arable farming that bore the brunt. Wheat growing contracted, whereas permanent pasture increased. Agricultural depression was widely experienced, particularly vulnerable areas being the heavy claylands of Essex and the area of light soils in Norfolk. But farming generally was hard hit and the landscape suffered accordingly: poor grade pasture returned, hedgerows were neglected, buildings fell into dilapidation, improvements were abandoned. The agricultural labourer fled the land.

The first decade of the twentieth century saw some recovery and a return to some comparative prosperity; unprofitability might remain the keynote, but there was at least a general stability in the overall picture. There was even growth in some areas: dairy farming fed urban demands, successful railway companies bringing vast quantities of milk to London and the big cities. By 1900 53 million gallons of milk were brought each year to London by rail, compared with 40 million gallons in 1890 and only 9 million in 1870 (Brown, 1987). As Essex was transformed from cereal farming to milk production, the Eastern Counties Railway pioneered the transport of country milk to London. The Great Western Railway (dubbed the Great Milky Railway) doubled its milk traffic between 1892 and 1910, and the LNWR's milk traffic grew by half between 1892 and 1904 (Perry, 1974). A growing sense of confidence among farmers led to the National Farmers' Union being formed in 1908. However, it was apparent by the time of the Great War that the country's agricultural industry had experienced a major trauma and there was no obvious recovery to former glories. The late nineteenth-century depression years for farming proved no passing wave, rather a permanent alteration in the country's economy.

With the country increasingly urban and industrial, the countryside seemed more and more residual: a land of small and medium-sized farms, largely tenant farmed, whose economic heart beat only fitfully. In some parts, even the landscape itself was in poor shape; 1891 was the peak year for crop and grassland in England and Wales (28 million acres), and the fall by over 3 per cent over the next two decades reflected the withdrawal of farming from the least profitable land, though urban encroachment for building land also played a part. H. Rider Haggard (1906), conducting his own survey of agriculture and social matters at the turn of the century noted that with more and more land being put to grass, fewer men were

The farming community before the Great War. Farm workers and visiting gentry, Alton Priors, Wiltshire in 1910. (By courtesy of the Rural History Centre, University of Reading)

needed for its working. 'Some parts of England are becoming almost as lonesome as the veld of Africa', he remarked (p. 540).

Industry

The face of parts of rural England and Wales had been severely affected during the economic transformations of the nineteenth century. The mining of progressively deeper coal in Northumberland and Durham, on both sides of the Pennines in Lancashire, Yorkshire, Derbyshire and Nottinghamshire, in the Midlands and in South Wales impacted severely on the countryside in those areas. The latest field to be exploited (Kent) began production just before 1914. New ironstone mining in the Cleveland Hills and Northamptonshire superseded non-ferrous mining elsewhere: copper in Cornwall and lead from the northern Pennines, areas which provided their own legacy of dereliction and abandoned waste. In North Wales the world's largest slate quarries (at Penrhyn and Dinorwic) reached the peak years of their production, while in Anglesey the remains of the Parys Mountain were but a reminder of the buoyant copper days of earlier years.

For the greater part of the countryside, the early twentieth-century manufacturing base remained the handicraft trades: hosiery in the East Midlands, lace manufacture throughout the Midlands, straw plaiting for hat-making in Bedfordshire and West Hertfordshire, the boot and shoe industry in Northamptonshire and a residual woollen industry in Wiltshire and Devon. But the key feature was contraction or extinction. For example, industries concerned with the making or processing of foodstuffs experienced major change, as a result of the development of large-scale facilities at deep-water ports. Windmills were progressively abandoned as milling was mechanized. Furthermore, the impact of industrialization had its effect on other traditional crafts. Machine joinery affected the wood worker's trade, the brewing industry turned to galvanized iron products, and everywhere low-cost, mass-produced articles hit rural crafts.

Features of the old, rural economy were systematically undermined; the manufacture of farming implements gravitated to engineering centres such as Ipswich and Lincoln. The blacksmith and the wheelwright retained their importance in meeting the needs of agriculture (the country was still essentially a horse-drawn society), but out-working to supplement the livelihood from agriculture fell victim to new employment patterns. By the beginning of the century any notion of a dual economy for rural England had passed: towns and cities were industrial and the countryside was agricultural.

Housing

Rural housing conditions at the beginning of the twentieth century mirrored the unequal social and economic structure of the time exemplified by the contrast between the agricultural labourer's cottage and the mansion of the landed gentry. At the bottom of the scale the market made it impossible for housing of any quality to be provided at rents that could be afforded. The farm worker was poorly paid and his employment often irregular; his living standards were significantly lower than those of his urban counterpart. The nineteenth century had begun with a legacy of inferior rural housing and, although enlightened landowners had made improvements, low wages throughout the period meant that house-building could never be profitable on any appreciable scale. Even when housing reformers and certain landed proprietors (notably the Dukes of Bedford, Devonshire and Northumberland (Darley, 1975; Havinden, 1981)) did improve dwellings, the accommodation was likely to be for permanent estate workers – a privileged section of the rural community.

By the outbreak of the Great War a little-changing situation obtained. The agricultural recession had done nothing to help, and the labourer's dwellings remained substandard, with deficient sanitary conditions. Determined steps to improve national housing had been taken in respect of urban areas, rather than rural, and the countryside had been virtually unaffected in the last quarter of the century by legislation designed to improve the lot of the poorly housed. Rural sanitary authorities had scarcely begun to use the powers available to them, in the Housing of Working Classes Act, 1890. The campaign for sanitary reform had been slow enough to take effect in the large towns, but it was even slower in the country towns and villages. By the last quarter of the nineteenth century some of the larger villages had water drainage installed, but questions of rural water supply, drainage and sanitation were still being raised in the 1930s (Mingay, 1990). Legislation in 1909 (Housing, Town Planning etc Act) led to surveys of rural housing being conducted, and some improvement was effected with regard to the closure of unfit dwellings, but the overall situation remained thoroughly unsatisfactory. An estimate by the Liberal Party in 1913 was that at least 120,000 new houses were needed in country districts.

Conversely, there were the grand, Victorian country houses, of which perhaps 2,000 were built – initially in a Gothic style but latterly in a vernacular tradition under the influence of the architect Norman Shaw. Set in their private grounds, well treed and landscaped with shrubberies and gardens (a taste for gardening was to pass down from upper to middle class), they represented a distinctive element in rural housing stock. The really great houses such as Cliveden, Mentmore and Eaton Hall could scarcely be emulated, but the successful middle classes maintained a search for country living, though in much more modest estates.

However, circumstances at the turn of the century were already suggesting that the extremes of wealth and privilege, reflected in gross housing disparities, might soon narrow. New housing styles were being mooted for different occupancy groups. A specialized dwelling, in the form of the bungalow, imported from India, had already made its appearance as a specialized vacation house within easy reach of London, largely on the Kent coast (King, 1984). After 1890 it became a much more widespread middle-class residence in outer suburban and country locations.

But the dominant influence of the time was represented by the vernacular traditions of the English cottage, and within a short period of time this common unit of design was applied to housing in both rural and urban areas. The new trend in design and spatial layout was powerfully promulgated by two remarkable men: Raymond Unwin and his cousin-in-law, Barry Parker. Their book, *The Art of Building a Home* (1901), showed how the new design principles could be applied to the layout of hamlets and villages, and Unwin's Fabian Society pamphlet, *Cottage Plans and Common Sense* (1902), continued the theme of the cottage and the community, linked by social purpose. This new approach to residential architecture was ideally suited to British circumstances, and it came to dominate the form of building development in the country as a whole for almost the first half of the century.

Even so, however, cottages were not necessarily cheap. The purpose of the Cheap Cottage Exhibition at Letchworth in 1905 was to ascertain whether a satisfactory cottage could be built for £150 (assuming a maximum rent of 3s per week, or rather less than £8 per year, to be paid by a rural labourer). In fact, the winning design (by Percy Houfton) cost £250. The fact was that cottage architecture, as a consistent thread in the form of new residential development for the 'lower classes', became uniformly applied to housing schemes which benefited the skilled artisan and the widening middle classes. This could already be seen at Cadbury's Bournville begun (effectively) in 1895 and at Rowntree's Earswick, laid out by Unwin and Parker in 1903. Hampstead was altogether a grander example, but in the years before the Great War the countryside saw an increasing number of cottage-inspired, romantic layouts in the form of a garden city (Letchworth), innumerable garden suburbs, a variety of copartnership housing schemes, also largely suburban in setting, and new colliery settlements (as at Woodlands, north of Doncaster, laid out between 1907 and 1912). It was really rather ironic that the humble, insanitary, agricultural labourer's cottage provided a prototype for a dwelling in the biggest expansion of mass housing England had ever seen – development which took place on rural land as towns and cities spilled over into adjoining fields.

The countryside was also being invaded by another urban group; residents who created their own makeshift landscapes of 'plotlands' (Hardy and Ward, 1984). Settlements of shacks and shanties colonized localities along the south and east coast of England; in south Essex self-built cottages were so numerous that they appeared as a vast rural slum (to be cleared and the area developed as a New Town – Basildon – in later years). These areas (and others elsewhere in the country) formed informal retreats for freedom-loving town dwellers. The plots were bought very cheaply, indeed sometimes acquired through squatting, and a new form of urban life was implanted with virtually no restriction on a countryside almost powerless either to resist the new incursion or to withstand the social and environmental consequences.

Other Urban Demands

By the turn of the century it was clear that urban growth was capable of making insatiable demands on the countryside; rural land was being taken up for urban requirements to such an extent that the country's land use pattern was irrevocably transformed. For example, increasing amounts of land were required to service a built-up area: sites were sought for refuse disposal for sewage treatment works and water purification plants, for public abattoirs to replace unhygienic, badly-located slaughter houses, for public cemeteries as old city churchyards became overcrowded and were closed, and for isolation hospitals to combat the spread of infectious diseases. The urban fringe of the late Victorian/Edwardian city became a zone of increasing competition in which suitable locations for urban uses, with their ever expanding space requirements, had to be found – a feature in even greater evidence throughout the twentieth century.

The urban areas were expanding territorially at an ever increasing rate, and land take was entirely at the expense of the rural periphery. In London's case the outward march seemed unstoppable. Late Victorian London was by far the largest city in the world and its continued growth caused consternation. On the one hand there was pride in size and power, but on the other there was fear at the threat posed by such a concentration of people over such a large area. Quite apart from residential needs, the outer periphery of built-up areas was also the location for new industry, for recreation grounds and public parks, for the new generation of golf courses, and for huge acquisitions of land required for the construction of docks and quays in the growing maritime ports which were catering for ever larger merchant ships.

Further afield the growing towns and cities had other urban demands. In the field of leisure, the late nineteenth century had seen the creation of a number of race courses around London and in the Midlands. Another type of race track was established at Brooklands, Surrey, in 1907 for cars. A long standing demand was for building materials; gravel, limestone and clay, roadstone and slate in geologically-determined areas. Brick making came into its own in the second half of the nineteenth century and spectacular demands for land were experienced in Bedfordshire as that county emerged as a major brick-making centre, with huge extractions of gault clay. The most extensive developments, however, took place in the working of the lower Oxford clay in the Fletton area, south of Peterborough from the 1880s; the first two Fletton brickworks were opened in 1897, and by 1910 clay extraction on a vast scale was permitting the manufacture of 48 million bricks a year (Cox, 1979).

Finally we should note the use made of rural land for urban water supply through large-scale reservoir schemes. Enormous increases in demand were being experienced by the end of the nineteenth century in response to rising population numbers in the city supply areas; also in the more extensive use of water in general as for industrial processes and domestic hygiene. Impounding schemes could be at considerable distances from the urban areas concerned: Manchester had already turned to the Lake District, and Birmingham to central Wales.

Local Government

Rural England and Wales at the opening of the twentieth century possessed a recognized social order of long standing, based on the possession of land. Throughout the previous century the aristocracy and the new gentry constituted an impressive pinnacle of power in countryside affairs, their estates efficiently managed by powerful land agents. By and large this established social order was maintained, though not without some conflict between land owner and the rest of the rural community over such matters as shooting interests, the game laws and poaching. A distinctive situation obtained in Wales where the land owners were anglicized, Anglican and Tory, while their tenants were Welsh-speaking, non-conformist and Liberal; hence rural grievances fuelled a radical, Welsh political programme. However, overall there was a settled order of a kind.

But this structure was changing; attacks on the monopolies of rural power were being mounted by political parties, agricultural trade unionism and other interests. In part the weakening of the monopoly resulted from the declining wealth (actual for some, relative for more) of many landowners. When agricultural prices collapsed from the late 1870s landowners ceased to comprise the majority of the really wealthy. As Moore (1981) puts it: 'Increasingly, this category was filled by men who left their money where they made it, in financial and industrial enterprises, who did not invest it in land' (p. 367). The importance of landed estates declined as agencies of social organization, and political power weakened – by no means quickly, indeed quite slowly, but inexorably.

This provided the background for the important developments which affected local government at the close of the nineteenth century. The system of health administration established under the Public Health Act, 1872 made the Poor Law Guardians the sanitary authorities in the rural areas. The Local Government Act, 1894 had adopted these areas as the new Rural District Councils, but now the authorities were based on direct elections. At the same time the parish was revived. By the beginning of the twentieth century, therefore, all the now familiar features of rural local government were in place: counties, districts and parishes with councils elected on a widening franchise, though with a still restricted male electorate. But there were oddities and weaknesses. Keith-Lucas and Richards (1978) remind us that the Rural Districts were essentially the poor law unions of parishes which had generally been formed on a five-mile radius pattern around a market town. The town had often acquired a charter or Urban District status. A typical Rural District therefore regularly had its urban centre omitted, with the consequence that the new local government boundaries were ill-matched to the needs of a unified area. The parishes, the smallest units of government, were given a new life within the Rural Districts. It was anticipated that they would counterbalance the influence exercised by the Church and the landowner, but the powers conferred upon them fell far short of establishing smallholders and peasant proprietors as a new yeomanry. Traditional power structures were eroded, but proved slow to change.

A TIME OF CHANGE

The essential features of rural England and Wales at the turn of the century have been sketched; 1914 has been suggested as a cut-off date, after which a number of circumstances served to invigorate certain directions of change. By the first decade of the twentieth century a number of critical characteristics of the countryside had become established, and in certain instances they were quite radical departures from the past. England and Wales were now decisively urban countries (indeed on population criteria they had been since 1851). Their wealth, based on manufacturing industry, was urban-centred, and the countryside seemed in decline. There had been a transformation from an agrarian rural society to one which was urban, and based on manufacture and trade. Agriculture, hard hit by foreign competition, was in poor shape and its manpower was falling. Land ownership, formerly the passport to social and economic power, now seemed less attractive. Rural trades could not compete with urban manufacture. The economic transformations seemed complete: a hundred years earlier the countryside had been the seat of both farming and local manufacturing enterprise, whereas the dual economy was now broken, agriculture spatially separate from industry. The urban dominance over the countryside appeared total: rural land was there to serve urban needs – to supply water and raw materials, to provide opportunities for recreation and to seek out new areas in which to live. And in the meantime, while cheap food could be imported, village communities stagnated.

But even as such a broad picture was emerging, portents of change were already suggesting the ways in which it might dissolve. By the beginning of the twentieth century a number of pointers could be detected, the full significance of which appeared in later years: a political dynamic which would affect landed interests, the development of a powerful nostalgia for a rural tradition, the effect of an anti-urban outlook on rural affairs, the adoption of rural conservation in the spectrum of middle class values and the consequence of changing technology in the revolution of road transport. All these five aspects were in place at the beginning of our period, and in fact they became dominant threads in the evolving story of the twentieth century.

Political Challenge to Landed Interests

We have already referred to the entrenched social and political power structures, based on land. From the last quarter of the nineteenth century onward, political agitation revolved round the struggle to redistribute the privileges which accompanied this land ownership. This complex story can only be touched on in passing, but wider reference is available in Douglas (1976), Horn (1976, 1984), Offer (1981) and Newby (1987).

The inherited hierarchy was formed by the landed aristocracy, whose general wealth was declining from the 1870s; the gentry, rapidly emerging in privilege in the nineteenth century, but with wealth derived from sources other than land; the landlords and tenants; and the labourers. In 1872 a national survey of land ownership was carried out, promoted initially by Lord Derby who confidently expected that it would confound the allegations of Liberal reformists that Britain was 'owned' by an elite of 30,000 landowners. The results proved embarrassing for the instigator. It revealed that about 400 peers and peeresses owned estates in England and Wales which averaged over 14,000 acres; roughly 1,300 'great landowners' (other than peers) owned estates of less than 7,000 acres; 2,500 squires and lesser gentry owned estates averaging more than 1,700 acres (Bateman, 1883). In Wales, more than 60 per cent of the land was owned by 571 great landowners, all with estates of more than 1,000 acres. Over 35,000 cottagers owned between them just 7,000 acres. Half of the county of Carnarvonshire was owned by landlords and a similar situation obtained in Merioneth (Jones, 1984).

In the later years of the nineteenth century and into the twentieth a powerful political attitude, expressed in Liberalism, moved towards a more collectivist view of the countryside. Rural political transformation took place with the extension of the franchise to male householders, which offered the possibility of diminishing the farming element in local politics, but in fact attitudes and power structures were slow to change. More was to be done in national politics and the 'Land Question' came to dominate political sentiment from the 1880s; the opposing creeds of Conservatism (which championed rate payers, agriculture and landowners) and Liberalism (which argued that the urban working class bore the brunt of local taxation) were sharply drawn. Henry George, the American economist and social observer, promoted the idea of a tax on land values in his book *Progress and Poverty* (1880), and his teaching became very popular.

Circumstances conspired to create uncertainty. The notion of redistributing economic power through peasant proprietorship in land was mooted, such as by the creation of small holdings, by local authorities. In 1887 local authorities were empowered to purchase land for allotments, and in 1891 County Councils were permitted to create small holdings of up to 50 acres.

The imposition of new death duties in the Budget of 1894 had shaken confidence in the security of land as a long term investment, just at a time when agriculture was severely depressed. The return of the Liberals to power in 1905 further raised the political stakes, with the Party's intentions to raise the lot of the tenant farmer, to improve the homes and working conditions of the labourer, and to attack the privilege and wealth of the aristocracy and gentry. Threats of higher taxation prompted a small group of estate owners in 1907 to form the Central (now the Country) Landowners Association. Land sales increased sharply in the years before 1914, perhaps 800,000 acres changing hands in a five-year period. The Liberal land campaign of 1912–14 advocated a minimum wage for agricultural labourers and the possibility of rent remissions from landowners. Meanwhile the question of small holdings had arisen again, legislation in 1907 and 1908 obliging County Councils to establish small holdings, if necessary through compulsory powers of land acquisition. Between 1908 and 1914 more than 14,000 holdings were created on some 200,000 acres, Norfolk being the most energetic county in this direction (Crouch and Ward, 1988). But the dream of a prosperous land-based yeomanry was still a long way off.

Lloyd George's Budget of April 1909 carried a number of radical new taxes, including those on land – for which a system of land valuation was necessary – a rise in death duties and income tax, and the introduction of super tax. In fact the land tax programme failed, but the information derived from the valuation process (for which a valuation office was formed as a branch of the Inland Revenue) provided a mass of information on land ownership, property boundaries and land values.

Change would not come quickly, or perhaps in expected ways, as later chapters will show, but an unstoppable pressure for reform had been applied. It seemed likely that the State, both central and local, would continue to engage with increased determination in progressive action to alleviate the position of the underprivileged. The Great War soon intervened, but not before first the influential book by the Hammonds', *Village Labourer*, was published in 1911 and then the authoritative Report of the Land Enquiry Committee had given its findings (1913). The Chairman of the Committee, Dyke Acland, concluded in a wide-ranging Introduction that 'The essential thing, unless we are to hold that the State is powerless to help, is that something effective should be done without further procrastination, for both the wages question and the housing question must be marked "urgent" in a large part of rural England today' (p. ix). The nub of the matter was that for the labourer 'his food is insufficient and his housing below the standard required either by decency or comfort' (p. xxiv) and that while 'no reforms of village life can succeed which are not founded upon a sound economic basis . . . no plea of economic necessity can justify conditions which are plainly indefensible from a humane and national point of view' (p. xxv). The challenge for reform was laid down.

Rural Nostalgia

Another broad perspective proved of long standing significance. Described as 'the pastoral impulse' (Marsh 1982), it represented a back-to-the-land movement. The inspiration was a deep-rooted anti-urbanism and anti-industrialism, of which the latter decades of the nineteenth century threw up a number of influential advocates. John Ruskin had a persuasive vision of a quasi-feudal agrarian society. William Morris' *News From Nowhere* (1891) was a fictional reconstruction of London devoid of ugliness and unwelcome manufacture. Edward Carpenter favoured the simple life style in harmony with nature; and the journalist Robert Blatchford in his book *Merrie England* (1894) (which sold over a million copies) also found a return to country living imperative. The urbanite responded to this deepening interest in, and love for the countryside, and rural England (especially) became a fashionable venue for new leisure pursuits including walking and cycling.

The pastoral impulse found expression in the Arts and Crafts movement, the revival of handicrafts marking a return to individual craft skills. The country was the only place it could flourish, pictorially 'epitomised in the drawings of Kate Greenaway, with demure country maids in bonnets and bows, set in neat gardens or rustic interiors, or in the graphic work of Walter Crane, where slim girls in flowing garments dance over windy meadows, symbolising spring and socialism' (Marsh, 1982, p. 144). William Morris moved his firm from central London to a mill on the River Wandle, south of London. C.R. Ashbee's Guild of Handicraft moved to Chipping Campden (Crawford, 1985). Furniture making was established in the Cotswolds and linen making again flourished in the Lake District. They were all small ventures but they were symptomatic of a yearning for a return to rurality.

Indeed, rural nostalgia began to colour many aspects of life. Health and a simpler lifestyle were particular considerations affecting tastes in, for example, dress and food. Heavy, confining Victorian dress was rejected in favour of simpler garments and woollen clothes were 'healthy' – as advocated by Dr Gustav Jaeger of Stuttgart. Sandals were adopted. Vegetarianism flourished as an antidote to a heavy Victorian diet; from Dr J.H. Kellogg's vegetarian sanatorium in Michigan, cornflakes became a world food product.

Country air was seen as the basis for a new healthy life. In well-to-do house architecture bedroom balconies were incorporated for open air sleep. Camping under canvas became popular, and the scouts and church brigades both embraced this tradition. By the beginning of the twentieth century there were camps run commercially, at Llandudno and the Isle of Man. Within a generation a huge expansion in organized camping would take place as one of the major outdoor, leisure-based activities of the century.

Town and Country

H.G. Wells' *Anticipations* (1902) envisaged a future London extending up to 70 miles from the centre (90 years on, he was not far wrong!). Certainly, the years at the turn of the century produced ample evidence that the capital and the provincial cities were expanding in territorial size at a rate faster than population growth itself should indicate. London was absorbing its peripheral villages and townships and the term 'Greater London' was already in census use; population at the centre was declining and for the last three decades of the century population increase had been confined to suburban rings. In the provinces change was less dramatic but trends were clear enough. In 1911 the boundary change for the City of Birmingham was a vivid institutional response to geographic reality: Parliament conferred sweeping additions comprising Aston and a huge suburban and rural fringe, which tripled the land area of the city.

There was a love–hate relationship with these huge agglomerations. On the one hand there was pride in London, the biggest city the world had ever known and the urban pinnacle of imperial power, and there was a certain awe in the economic muscle of the manufacturing cities, the earliest and largest in an industrializing world. But hand-in-hand went the perception of urban degeneration. The argument was that big cities were not 'natural' for people, they were unhealthy, living conditions for many were abject and they were centres for crime. A moral view took hold that the de-humanized urban concentrations should be rejected; by comparison, as we have seen, the virtues of rural England were increasingly acknowledged.

Trends in population distribution were already reflecting a looser housing spread, the suburban railway in London, at least, assisting in an ever decentralized locational pattern. Suburban living (increasingly low density in the form of garden suburbs) became fashionable and the practice spread down the social scale as soon as economic circumstances permitted. But there were other ways in which the poor and deprived Londoner might be helped to escape from the polluted and corrupting city. Philanthropic factory owners saw both social and economic sense in providing spacious, improved housing conditions for their workers (and others) in rural surroundings: at the turn of the century Port Sunlight, Bournville and New

Earswick were prime examples in providing new settings for a contented, healthy workforce. Housing reformers generally urged the advantages to be derived from the new forms of housing design and estate layout. Henrietta Barnett's Hampstead Garden suburb enabled Raymond Unwin to create a residential setting in which houses were grouped around green spaces.

Local authorities themselves were swept along in the direction of building at lower densities. The housing architects of the LCC impressively caught the tide in the out-of-town estates before the Great War. Manchester Corporation followed suit with the Blackley Estate to complement other developments at Burnage Garden Village, the private estate at Chorltonville and the co-partnership project of Fairfield Tenants Ltd (Harrison, 1981).

The new garden city movement, based on the advocacy contained in Ebenezer Howard's book *Tomorrow: a peaceful path to real reform* (1898), reissued as *Garden Cities of Tomorrow* (1902), was an even more explicit rejection of the old city. Letchworth, dating from 1904, was created to demonstrate how the city and the country could be 'married' in a strategic solution which saw clusters of garden cities forming 'social cities' as satellites at a distance from the parent, with an agricultural belt between.

These examples of how urban form was progressively loosening demonstrate the widespread conviction that it was both necessary (on health grounds), desirable (environmentally) and financially prudent (bearing in mind land costs) to live at lower densities in keeping with new design fashions. The norm of 12 to the acre may not yet have become standard practice, but Unwin's pamphlet in 1912 'Nothing Gained by Overcrowding!' proclaimed a general intent. It was in accord with an increasing view that the best way to solve the housing crisis of the day (overcrowding, high densities, squalor and insanitary conditions) was to leave the city and build afresh on cheap land in the periphery. In the years before the Great War this pointer to future trends was powerfully demonstrated.

For some, the flight from the city was accompanied by a return to farming enterprises. The direct transfer of overcrowded town dwellers to rural locations was often advocated. A conscious resettlement programme, for example, was sought by General William Booth of the Salvation Army, in the form of colonies. Agrarian communes had been quite a feature of the nineteenth century (Hardy, 1979) and the promise of spade husbandry resurfaced as a solution to the problem of urban unemployment, which increased particularly in London at the turn of the century. In Essex Booth opened a colony near Southend and the Poor Law Guardians from Poplar established settlements first in south Essex and later in Suffolk. The farm colony movement withered and did not survive, yet the issue of a return to the land would recur in different guises later in the century.

Rural Conservation

The years marking the turn of the century threw up a number of contentious issues which illustrate a growing interest in the protection and presentation of a rural landscape heritage, long under threat both from countless instances of urban development and a widespread apathy about a cultural past. By 1914 the agenda was a rich and varied one.

The most long standing issue concerned the enclosure of the commons. For half a century and more, open common land especially around London had become the subject of urban encroachment and dispossession for private development. The Commons Preservation Society, founded in 1865, with purposes to secure 'the use and enjoyment of the public open spaces, situate in the neighbourhood of towns, and especially of London, still remaining unbuilt upon', proved very influential (Williams, 1965). Over the years it was successful in opposing building on Hampstead Heath, and it was a rallying point in saving Wimbledon Common and Putney Heath. Other London commons saved from development included Plumstead, Coulsdon, Banstead and Petersham. Elsewhere, Berkhampstead Common and Epping Forest proved *causes célèbres*, their defence marked by violent public agitation.

The National Footpath Preservation Society, founded in 1884, merged with the Commons Preservation Society in 1899; 'Open Spaces' was added to the title of the joint society in 1910. An increasing problem was access to the countryside, and in 1905 a number of walking groups amalgamated in the Federation of Rambling Clubs – from which sprang the Ramblers' Association, formed much later in 1935.

Buildings too required action for their preservation. William Morris had been instrumental in the founding of the Society for the Protection of Ancient Buildings in 1877, the flashpoint being proposals announced for the reconstruction of Tewkesbury Abbey. The Rev Hardwicke Rawnsley, Canon of Carlisle, launched a separate initiative in association with Octavia Hill and Sir Robert Hunter in 1895. They founded the National Trust for Places of Historic Interest or National Beauty for the purposes of acquiring land and buildings by gift or purchase in order to preserve them.

The questions of public access to, and an enjoyment of the countryside were central to these various proposals. Accessibility subsequently proved to be a lively issue to the present day. For example, successive Access to Mountains Bills, seeking to give the public a right (albeit restricted) to walk on specified private property without being subject to the law of trespass, all failed, but out of these early steps an embryo National Parks movement gathered momentum.

Another initiative led ultimately to the green belt. Reginald Brabazon, Earl of Meath, elected an Alderman by the first London County Council, became the first chairman of the LCC Parks and Open Spaces Committee. He was an early advocate of a system of circular boulevards for London, linking parks and open spaces. In 1901 the idea was taken forward by William Bull, MP for Hammersmith, who recommended 'a continuous chain of verdure' around London as a memorial to Queen Victoria (Aalen, 1989). His 'green girdle' would form an oval ring around the outer rim of London. In 1911 George Pepler proposed a parkway around London in the form of a strip of land a quarter of a mile wide. In the meantime Ebenezer Howard's garden city (Howard 1898, 1902) envisaged an agricultural setting for the new townships, presaging an altogether wider view of urban fringe management.

Communications

At the very time when so much interest in the countryside and what it represented was being expressed, largely by the middle class urbanite, technical innovation in means of transport meant that increasing numbers of people could go further afield in a day's journey than ever before. In due time the sheer pressure of numbers of people flooding into the countryside and sampling its pleasures was to create a problem in its own right.

Cheap train travel transported workers from the industrial cities, and of course London, to the popular seaside resorts through landscapes of a character very different from their daily urban experiences. Possibly the most frequented inland resort area, served by train, was the Lake District where Windermere in particular, with its steamer trips, proved a popular destination. More independent travel was afforded by the bicycle, particularly after the modern 'safety' machine, with detachable pneumatic tyres, was introduced in the early 1890s. The travels of H.G. Wells' Mr Polly recounted in the novel of 1910 suggest the ready way in which the bicycle became the liberator for the Edwardian lower-middle class in week end travel. The Cyclists Touring Club, founded in 1878, soon formed a strong pressure group in countryside affairs; it reached a peak membership of 60,449 in 1899 (Patmore, 1970).

But it was the motor car and the advent of popular motoring which provided the most dramatic development. The number of licensed motor cars and cycles rose from fewer than 8,500 in 1904 to 90,000 in 1910 and nearly 230,000 at the outbreak of the Great War. The motoring public was still drawn from a narrow band of well-to-do people, but the beginnings of new environmental problems were quickly seen: road hazards, nuisance (dust and noise), and the cost of road improvement and maintenance.

Finally, we should note the arrival of the air-plane and its demands – one of the most astonishing developments of the twentieth century. From around 1910 intrepid pilots were landing their flimsy machines in grass fields on the outskirts of cities across the country. Within just a few years permanent sites were being prepared, an extra fillip provided by the demands of war.

3

THE HIATUS OF WAR 1914–1918

It was not the countryside that presented the nation with its major problems in 1914. For some years rural questions had been relatively low on the national political agenda, except for the simmering disputes over land and taxation. The worst of the agricultural depression of the 1890s was over and there had been a return to some stability. Instead, political attention was focused elsewhere: on Ireland and the Tory rebellion over Ulster; the rise of the suffragette movement; and the growing power of organised labour in strikes and worker unrest. However, with the outbreak of hostilities in August 1914 the world changed abruptly; at home a new set of issues brought the countryside to a fresh importance in national affairs.

'Rural England in the early summer of 1914 had an atmosphere of modest prosperity'; this is the judgment of Howkins (1991, p. 252). Horn (1984) also writes of the decade before 1914 as one of comparative prosperity, with a steady revival in the value of agricultural output, and of rent receipts. Earlier, contemporary observers had expressed varied assessments. The social and economic historians, J.L. and Barbara Hammond (1911), stressed the many hardships of the village labourer during the nineteenth century, but R.E. Prothero (later Lord Ernle), writing about English farming in 1912, pointed to improved social conditions for the labourer. That year the agricultural scientist A.D. Hall, (1913) had finished his three tours through England; he would also write optimistically of the state of the country's farming. We might simply conclude that, as we have seen in the previous chapter, rural affairs were undergoing profound change, unfinished revolutions of the last century still working their way through social, economic and political relations. Flora Thompson later recorded hamlet, village and country town in her trilogy *Lark Rise* (1939), *Over to Candleford* (1941) and *Candleford Green* (1943), subsequently brought together in one volume *Lark Rise to Candleford* (1945). In the words of H.J. Massingham (1945), who wrote an introduction to the single volume she revealed 'a local self-acting society living by a fixed pattern of behaviour and with its roots warmly bedded in the soil.' But the pattern was disintegrating and the roots were loosening. The old order of rural England typified in the hamlet of Lark Rise had already gone in the suburbia of Candleford Green (Thompson's depiction of Buckingham).

We can detect both the same nostalgia and the resignation to the inevitability of loss with George Bourne's *Change in the Village* (1912). Bourne took his pen name from the Surrey heathland village, where he lived, to the south of Farnham. He regarded The Bourne as his laboratory, a community which he subjected to sustained observation over a period of twenty years. He concluded that the rural English were between two civilizations: one lapsed, the other in the process of being formed. He caught the sense of transition:

> The old life is being swiftly obliterated. The valley is passing out of the hands of its former inhabitants. They are being crowded into corners, and are becoming as aliens in their own home; they are receding before new comers with new ideas, and, greatest change of all, they are yielding to the dominion of new ideas themselves . . . A lath and plaster partition may separate people who are half a century asunder in civilization . . . in another ten years' time there will be not much left of the traditional life whose crumbling away I have been witnessing. (p. 11)

The rural life in the Victorian and Edwardian countryside which, *inter alia*, Pamela Horn (1976) and Sadie Ward (1982) describe, had many undesirable features. Yet when war came in the summer of 1914 the countryside evoked strong patriotic feelings; Dakers (1987) points to three writers (Siegfried Sassoon, Rupert Brooke and Edward Thomas) whose sentiments surfaced strongly at this time to proclaim the exceptional, perhaps mystical qualities of rural England. The irony was that the countryside which inspired them to fight in its defence would prove only too vulnerable to change, more rapid and fundamental than their generation could possibly have anticipated. Consider:

> Britain was the only European belligerent to increase her supplies of home-produced food but to do this three million acres of grassland were ploughed up. Thousands of acres were requisitioned for army and prison camps, munitions' factories and airfields; extensive areas of woodland were devastated to provide extra timber for use at home and abroad; some country towns and villages were filled to overflowing with officers, soldiers and the machinery of war; country houses were turned into hospitals and their parks were churned up by gun-carriages and tanks; in the fields labourers were replaced by truant school children, land girls in breeches, able-bodied pacifists, Belgian refugees and German prisoners. (Dakers, 1987, p. 17)

In this chapter we outline the various ways in which the Great War impacted on rural England and Wales: with regard to labour, land and food supply and to changes in village and community life; and in respect of housing and planning. We describe the four-year period as a 'hiatus', and it is the case that there was a break from the past, but in some respects, notably in housing and planning matters, the years were marked more by continuity from the past. Moreover, while the consequences of war wrought profound change, they set in train a number of developments which took some time to unfold. Hence, although this chapter is concerned essentially with the period 1914–18, in order to 'round off' the particular story it is necessary to stray a little into certain events of the years after 1918. This will become clear in later pages.

LABOUR

Lord Kitchener's appeal shortly after the outbreak of war for 100,000 new recruits was met with a speedy response – even to the sudden departure of labourers at harvest time. The landed classes lent ready encouragement to their work force to enlist, the nobility showing a traditional zeal in this respect as organizers of local militia. By the end of 1914 nearly 1.2 million Britons had volunteered, including substantial numbers of young country

The wartime agricultural labour shortage. German prisoners of war bagging potatoes on a farm in Surrey towards the end of the Great War. (By courtesy of the Rural History Centre, University of Reading)

men. Eventually perhaps about a quarter of a million men were recruited from farming to serve in the Armed Forces or in other areas of war production. Contemporary figures of numbers recruited have been the subject of controversy, Dewey (1989) arguing that figures were greatly exaggerated. Certainly, however, recruitment amongst the unskilled, as opposed to the skilled, was more pronounced.

In any event a labour shortage ensued. By 1916 the conventional labour force had fallen to 91 per cent of the pre-war level and subsequently declined to 89 per cent – a shortfall intensified, of course, by a 'ploughing up' policy, since arable land required considerably more labour for a given acreage than did pasture. Labour supply became acute and arrangements had to be made for temporary releases of military personnel to return to farming at busy times of the season; 84,000 soldiers were released by the end of the war. Meanwhile there were growing numbers of killed and injured men who would never return to their former occupations. It is a sobering statistic that nearly one in eight serving in the Armed Forces during the war was killed.

The labour crisis obliged the authorities to turn elsewhere for labour replenishment and increasing efforts were made to attract women. Prior to the Great War the number of women in agriculture had been falling; neither was there any immediate recovery as better paid jobs in the service trades became available. Importantly, there was a separation allowance for married women, and family income could often be supplemented from providing billets for the local military. Recruiting women to the land required considerable nurturing through campaigns which involved a call to patriotism. The Women's National Land Service Corps (WNLS) gave momentum to calls for women's employment, but finally it was the Women's Land Army, formed in January 1917, that succeeded both in improving the training and skill of recruits and in increasing numbers to the peak of 16,000 in September 1918. It is estimated that 300,000 village women were working on the land in 1918, the great majority part-time; two-thirds of these represented the war-time increase (Mingay, 1990).

There were other types of substitute labour. The number of German prisoners of war increased from around 5,000 employed on the land at the beginning of 1918 to more than 30,000 by the end of the year; by and large they were good workers and their labour was valued. School children were also recruited. Education authorities passed bylaws to enable children to work on farms during school terms and holidays.

LAND AND FOOD SUPPLY

At the outbreak of war, of all the belligerents Britain was the most dependent on imported foodstuffs (Barnett, 1985). Its agricultural sector was relatively small and the country was importing 60 per cent of its total food supply (80 per cent in the case of wheat, 75 per cent of its fruit and 40 per cent of its meat). During the summer of 1915 German submarine attacks on British shipping heightened anxieties about maintaining the level of food imports. In a further blow, grain imports from the Black Sea areas of Russia, Turkey and Rumania (which had accounted for 35 per cent of the British total in 1911) were lost by the closure of the Dardenelles and Turkey's alliance with the Central Powers.

There had been no stockpiling before the war and food control assumed an importance far greater in Britain than in other countries. The government's policy with regard to food production first relied on exhortation to increase crop acreages and food supply, but this soon required reassessment. A special Departmental Committee, chaired by Viscount Milner, was appointed in June 1915 to consider the new situation, with particular reference to future harvests. In a speedily produced Interim Report (in July), the Committee took the view that farmers required the security of a guaranteed price for wheat of 45 shillings per quarter (or 1p per kilo expressed in decimal currency and metric units) for a minimum of four years, if they were to be persuaded to increase wheat production. But the Cabinet declined to support the recommendation and the Committee's final Report (in October) did not take that matter further.

However, the 1916 harvest was both smaller in extent and poorer in quality than that for 1915. Moreover, food prices rose as a consequence of intensified submarine attacks. Demands for government action led first to milk prices being controlled and then to arrangements for fixed maximum prices, extending to a variety of agricultural products. A system of price controls replaced *laissez-faire* and, with Lloyd George appointed Prime Minister in December 1916, an altogether more vigorous policy was established. Executive Committees of the County War Agricultural Committees were set up, charged with deciding on land usage and efficiency in cultivation and the type of crops to be grown; ultimately their responsibilities extended to labour supply and provision of machinery and imported fertilizers and feeding stuffs.

A 'plough-up' campaign sought to increase the amount of arable land in the country and reverse the predominantly livestock regime that had marked the years before the onset of war. The work of the Executive Committees came to represent a demonstrably dirigiste farming policy, stimulating the cultivation of desired cereal crops, regulating land stocking, and bringing back into cultivation poorly managed farm land. The Committees were used to enforce on farmers a major shift in production, central to which was the ploughing up of pasture for grain and potatoes. It was calculated that 100 acres of pasture would feed only nine persons if used for fattening, and 41 persons if devoted to dairying. Ploughed up, that same 100 acres would maintain 172 persons if planted with oats, 208 if planted with wheat, and 418 if planted with potatoes (Mingay, 1990). The ploughing up campaign added only marginally to the harvest of 1917. The effects on the harvest of 1918 were much more substantial, however, with the acreages of wheat and oats up by one-third over the 1916 total and potatoes up by one quarter. By the end of the war the overall area of pasture had fallen by 9 per cent, bringing it back to the acreage of 1890. By 1918 British agriculture was feeding the nation's population for a proportion of the year 25 per cent greater than in 1914 (155 days, as opposed to 125 days).

Interventionism became fully explicit with the Corn Production Bill, introduced to Parliament in February 1917; it received the Royal Assent in August. The new President of the Board of Agriculture, R.E. Prothero, who had been a member of the Milner Committee, was a supporter of the system of guaranteed prices. His Act returned to this theme and also dealt with other long-standing farming questions. Guaranteed minimum prices were offered for wheat and oats. A Central Wages Board was established to deal with the question of workers' wages, and a national minimum wage of 25 shillings per week was established. Farm tenants were protected from rent increases. Finally, enforcement powers in respect of land cultivation and stocking, and the issuing of compulsory cropping orders were provided.

Barnett (1985) suggests that it was formerly asserted that food supply in the event of war was not an issue that attracted much interest before 1914, and that it was only because of much more strenuous action after 1916 that the government successfully overcome the national food crisis. Her revisionist interpretation argues that while it is true to say that agriculture was not exactly put on a war footing in the early period, and that Lord Selborne was an ineffective President of the Board of Agriculture, nonetheless certain action by the two war governments led by H.H. Asquith (the Liberal government in office when war broke out and the coalition government that replaced it in May 1915) did lay important foundations for the controls introduced in 1917 and 1918. A central administrative unit, destined to form the core of the Ministry of Food in December 1916, was already working on issues like regulation of prices and distribution, and in conservation.

The introduction of bureaucratic interference, particularly through the work of the Executive Committees, was adjudged necessary in wartime conditions. The desperate food situation no doubt justified a 'stick and carrot' system of cajoling farmers to act in the national interest. In 1917 and 1918 farmers increasingly warmed to the idea of government price cushions. Lloyd George's pledge to maintain guaranteed grain prices assured him of the farming vote in the post-war elections. But the arguments between continued state regulation of agriculture and a return to a reliance on farmers' own inclinations were finely drawn, on both sides.

In the event, new circumstances determined the outcome, at least for some years. To round off the story, we may summarize the events as follows. The guarantee system was retained for a while after the war, under the Agriculture Act of 1920, which reaffirmed the financial security afforded by the Corn Production Act, 1917. But a collapse of grain prices in only the following year threatened the State for the first time with large compulsory payments. Government repudiated the agricultural subsidies when the Act was repealed; guarantees were terminated and, for the workers, protection of the Wages Board was lost. The Ministry of Food was wound down and finally closed in 1921. Once again the country's agriculture faced an uncertain future, open to the vagaries of world markets, with recently-won arrangements for state intervention abruptly withdrawn.

VILLAGE AND COMMUNITY LIFE

The four years of war had a considerable impact on rural communities. We have already referred to changes in the labour force: the withdrawal of substantial numbers of men working on the land, in many cases never to return, and the partial replacement by women, children, temporary soldiery and prisoners of war. But there were other adjustments to make. Army camps, both permanent and temporary, impacted on village life; men had to be billeted and new, sometimes well paid, jobs were suddenly available in camp construction and a variety of service trades. Aerodromes were constructed and munitions factories built; property was requisitioned from the earliest days of the war. Craft workers found new jobs, thus creating a shortage of skilled labour. Military demands seemed insatiable, even down to the number of horses which were required in massive numbers. The introduction of tractors to supplement horses seemed to confirm that an old order was changing.

Heavy fatality figures at the Front compounded the sense of disruption and instability. 'The English countryside became a memorial to the dead,' as Dakers (1987, p. 17) puts it. Village war memorials tell their own story of personal and community loss. Farmers lost sons and their ablest, strongest labourers. Deaths among the upper classes had particular consequences as country estates were left without heirs. The burden of death duties levied on the estates of those killed in action was reduced by virtue of special legislation, but could still be substantial. With 'double deaths' – an estate owner followed by that of his heir – death duties could necessitate the disposal of property.

For one reason or another land ownership appeared less and less attractive. The pre-war, Edwardian intent to liquidate the landed interest, which was defused by the war, returned after the war's end with a vengeance. An estate which had paid 9 per cent of its gross rents in income tax, land tax and rates before 1914, might be paying 30 per cent in 1919 (Dakers, 1987). The Budget of 1919 introduced death duties of 40 per cent (raised from 20 per cent) on estates valued at £2 millions and over. Rents had remained steady for years; many landowners preferred not to raise them, but to sell instead.

Again, it is necessary to complete the story. An avalanche of sales, beginning in 1919, had the effect that by 1921 perhaps one-quarter of the land of England (some six to eight million acres) had changed hands as estates were broken up. In most cases title passed to sitting tenants, thereby significantly extending the owner occupied sector in the country's farming pattern. Country houses came onto the market and, as the landed gentry decamped to the city or to dower houses, new uses were introduced to their residences – private schools, hospitals and mental asylums, as examples.

But war-time and post-war disruption was accompanied by attempts to promote solidarity and new relationships. Of these the most enduring was the growth of the Women's Institute movement (Jenkins, 1953). Originating in Canada in 1897, the first Institute opened in Anglesey in 1915. There were forty in existence by the beginning of 1917 and the forming of new Institutes became the responsibility of the Women's Branch of the Board of Agriculture. By October 1919 there were more than 1,200 Institutes when the Board handed over responsibility to the movement's own National Federation under the chairmanship of Lady Denman. The WI had already emerged as a contributor to the regeneration of rural life, a role it has continued to discharge.

Attempts to resettle war combatants, particularly the disabled, often focused on village communities. Initiatives straddle the latter period of the war and the immediate post-war years. The government set up a committee to consider land settlement for ex-servicemen, chaired by Sir Harry Verney, a large landowner. Smallholdings Acts were passed in 1916 and 1918. The Land Settlement (Facilities) Act, 1919 included a special section applying to cottage holdings. Between the Armistice and 1924 County Councils bought and leased 250,000 acres for this purpose. Within two years of the end of the war more than 14,000 people had been settled on the land.

There were various private schemes which aimed to assist disabled soldiers and sailors to resettle in the country. The garden designer and town planner, Thomas Mawson, drew up a scheme for creating villages 'replete with industries, handicrafts and horticultural pursuits, for the shattered men thrown upon the nation as the aftermath of war' (Mawson,

1927) in the memory of his son who fell at Ypres in April 1915. Mawson envisaged a series of industrial villages financed from donations and public funds; an Industrial Villages Interim Committee was formed to further the scheme but a lukewarm government response led to only a small number of settlements being started (Cherry, Jordan and Kafkoula, 1993). Mawson was but one initiator. Another initiative was that by a body known as the Village Centres Council which purchased an entire village – Enham, Hampshire. Private philanthropy set up village homes in Shrivenham, Berkshire, and the Women's Scientific Cooperative Farm was located at Sutton Valence, Kent (Dakers, 1987).

DEVELOPMENTS IN HOUSING AND PLANNING

By 1914 the new planning movement had gathered pace and direction. The hiatus induced by war could have been fatal, but the inevitable disruption over four years proved not to be so. Sufficient happened during this period to sustain fledgeling developments, while particular circumstances in the closing years of the war served to give the protagonists in matters of housing and planning renewed vigour.

The permissive powers provided by the Housing, Town Planning etc Act, 1909, whereby local authorities might prepare Town Planning Schemes, remained in place, though shortages of manpower in local councils meant that work was soon sharply curtailed. But at least Schemes prepared before the war for parts of Birmingham and Ruislip-Northwood were going ahead, covering land which was semi-rural. Planning activity fell away after 1915 but it was still the case that by 1919 no less than 172 Schemes had been authorized to be prepared or adopted by local authorities in England and Wales, involving over 300,000 acres (Cherry, 1974). Given the renewed interest in town planning in the context of new housing and reconstruction after the war, the fact of a proven take-up by local councils since 1909 was important. The Town Planning Institute, founded in 1914, maintained its professional journal throughout the war years and its leading members were in a position to bring influence to bear on the reform movements of the day.

The first President of the new Institute, Thomas Adams, was Chief Town Planning Inspector to the Local Government Board. On leaving for an appointment in Canada, his successor, Raymond Unwin, followed him, both as President and at the Board. It was largely through Unwin that earlier initiatives in housing and planning were kept alive, and indeed advanced (Miller, 1992). At first it was through State housing experiments in connection with the production of munitions. The rapid expansion of the wartime munitions industry distorted housing supply and demand in a number of areas. The building of a large estate at Well Hall, Eltham, south east London, provided accommodation for the increased work force at the Woolwich Arsenal; the architect Frank Baines created a suburban environment of romantic, informal quality. The new residential fashions in layout, density and architectural treatment, seen before the war, were continued not only in England and Wales but also in Scotland. Unwin himself was drafted into housing work for the ordnance factories, and under his direction plans for two new townships in Scotland were drawn up in the largest of the wartime schemes. These were at Gretna and Eastriggs, north of the Solway in Dumfriesshire. Following his work at Letchworth and Hampstead these two schemes further elaborated his hold on the progressive ideas in housing and planning in the country at that time. The Great War did not break the continuity from the past in this respect.

This was fortunate because as early as 1916 plans for reconstruction after the cessation of hostilities were under discussion. Lloyd George's Reconstruction Committee, assembled in February 1917, was followed by a Ministry of Reconstruction headed by Christopher Addison. Support was given to the establishment of a Ministry of Health and a huge housing programme for the working classes (the implementation of which is described in chapter 4). Estimates varied, but clearly there was a need for large numbers of new dwellings to be built (between 120,000 and 400,000) and inevitably the majority of these would be built in urban-rural fringe localities.

Equally to the point was the type of dwelling to be built, and Unwin's low-density, cottage architecture instincts prevailed. During the war the Local Government Board sponsored a committee, chaired by Sir John Tudor Walters (a Director of the Hampstead Garden Suburb Trust), to consider building construction in connection with the provision of dwellings for the working classes. Their report, published in 1918, owed much to Unwin as a member, the layout and design proposals following his own prescriptions in schemes undertaken earlier in his career. The recommendations of the Tudor Walters Report (1918) were adopted in the *Housing Manual* of 1919 and were not overtaken for more than a quarter of a century; they established layout and design norms in the public sector (and by imitation in the private sector) throughout at least the next two decades. The proposals covered density, site planning and house design, and specified a high standard of working-class housing. In a growing mood of determination about the course of post-war reconstruction, political speeches on a 'homes for heroes' theme striking a popular chord, the prospect opened up of tackling the abject housing standards of the rural areas.

Meanwhile continuity was effected by the garden city movement, represented by its organ the Garden Cities and Town Planning Association (GCTPA), as it was then called. It was indeed just one period of a long campaign (Hardy, 1991). The Association was keen to keep the example of Letchworth before the public eye and its President, Lord Salisbury, chaired a specialist Housing Panel for Lloyd George's Reconstruction Committee (reactivated in March, 1917), so keeping alive an influential voice on housing matters. The Association advocated a land settlement programme for returning soldiers and seamen, with three experimental colonies, but the idea failed to find favour.

However, Howard's original notion of a garden city continued to attract, and a small group within the Association campaigned vigorously for new housing to be built on garden city principles after the war. A young wartime pacifist, who had been Secretary-Manager to the Howard Cottage Society at Letchworth, F.J. Osborn, was prevailed upon to publicize their cause: under his pseudonym 'New Townsmen', *New Towns After the War* was published in 1918, echoing Howard's ideas but now advocating new towns as part of a State programme. In fact, as we shall see, the reforming zeal was soon lost in the country and the Association's call for the State harness was not taken up, but importantly the protagonists' cause was stronger at the end of the war than at the beginning. Wartime reformist enthusiasm for a cause helped to set the agenda for the post-war years. The implications for housing in England and Wales over the next twenty years were profound; as we see in the next chapter, only one additional garden city (Welwyn) was built, but dispersal out of the congested cities to green field peripheries in estates of low density development went at least some way to meeting the Association's objectives.

4

THE INTER-WAR YEARS

Broadly speaking, given the unfinished business which spilled over from the end of the war into the early 1920s, the inter-war period was neatly self-contained, within the aftermath of one world war and the onset of another. The two decades were marked by renewed militarism in Europe and the Far East, and by uncertainty in the stability of the world capitalist order. The increasing threat of disruption to world peace caused by the growing power of totalitarian governments provided an uneasy background for Britain, plagued as it was by periodic bouts of social strife and unaccustomed jolts to advances in material well-being. The country's world role as an industrial and trading nation was under pressure as two economic recessions (1921 and 1929–32) were experienced, the effect of the second exacerbated by the US stock market crash of October 1929 and the international financial crisis of 1931.

During these twenty years Britain's major domestic problems related to the changing fortunes of the industrial heartlands which had formed the basis of Victorian prosperity – essentially the coalfields and the tidal estuaries and ports. These areas now underwent hard times experiencing high rates of unemployment, social distress and the makings of new regional distinctions between economically 'favoured' and 'unfavoured' districts. Their industries – mining, steel and iron manufacturing, heavy engineering, textiles and shipbuilding – faced severe competition. On the other hand, relative prosperity in England, notably in the southern parts of the West and East Midlands, the Home Counties and the other metropolitan fringes of London was supported by the growth of new 'light' industries, the buoyancy of car manufacture, the service trades and a range of consumer industries parasitic on the spending power of an increasingly home-oriented society. It should be recalled that London and South East England then constituted the largest single domestic consumer market in the world. A pronounced drift in population took place, overriding the long-established net rural-urban movement and the later inter- and intra-migration; this time it was north-south, responding to the new economic opportunities. Greater London increased its population by two million between the wars: 1¼ million of these were from inward migration (Cherry, 1988).

In this context the affairs of rural England and Wales might have been considered small beer. But urban and regional concerns impacted on the countryside in a number of different ways and the inter-war period proved to be significant in both establishing the agenda and charting paths for the future. The purpose of this chapter, in pursuing the chronological story, is to introduce a number of themes which became the subject of later experiences. Beginning with agriculture and the continuing concern with the land, it will be seen that wartime productivity was not maintained and depression returned. Second, with regard to matters of countryside recreation and the mass demands of the urbanite, developments here had already been foreshadowed, but the new scale of popular demand proved expansive. Third, concerning the urban pressures on rural land from the point of view of house building in the countryside, the inter-war period demonstrated a vigorous growth in low density suburbanisation. Next, consideration is given to the nature of the countryside lobby in response to these developments: on the one hand there was pressure for conservation and protection, while on the other there was pressure for access to land for recreation purposes. Fifth, the growth of forms of State regulation is examined, particularly with regard to land use planning, during this period. The 1930s (especially) saw increasing pressure for an interventionary role by the State, which later periods built upon. Finally there is a reflection on the sum of the rural issues at the close of the 1930s; across a period of twenty years there was inevitably both change and continuity, but in many ways the two decades might be regarded as formative for what transpired later.

AGRICULTURE AND THE LAND

In the previous chapter it will be recalled that in the Agriculture Act, 1920, the wartime experiment with guaranteed prices was continued, but following the slump in world prices, the Act was repealed in 1921. A new situation for farming unfolded, as described by Creasey and Ward (1984). In the 1920s Britain became a dumping ground for the world's agricultural surpluses. The wheat exporting countries raised their output, and chilled meat from the Argentine, meat and dairy produce from New Zealand, butter, bacon and eggs from Denmark and fruit from North America flooded the home market. British agriculture could not compete and an early casualty was the slashing of agricultural wages. A fall, on average, from about 40 shillings (£2.00) per week in 1920 to 25 shillings (£1.25) per week at the beginning of 1923 produced a widespread strike of labourers in

Norfolk. A bitter struggle led to no more than an empty victory for the labourers (it was agreed that there would be no fall below 25 shillings [£1.25]), with many strikers unable to return to their jobs.

The 1920s passed with little relief extended to the farming community. A return to subsidies came in the mid-1920s, but this was only on beet-sugar, to encourage a comparatively new crop. An Agricultural Mortgage Corporation, established in 1928 with limited public funds, perhaps promised more, and in the next year Chamberlain's Local Government Act abolished agricultural rates (a measure which particularly benefited landowners). But in fact there was worse to come. Following the collapse of the US stock market, world wheat prices fell by half between 1929 and 1931. Britain was unprotected; most European countries intensified their existing protectionist measures, and the volume of agricultural products entering Britain increased still further. With wheat and barley prices badly hit, farmers reduced their arable acreage and progressively abandoned marginal land to reduce labour and other costs. The conversion of 1,000 acres of arable to grass could save up to forty jobs. Inter-war farming change was precisely most marked in this transfer from arable to pasture: a wartime peak of 15.7 million arable acres in Britain fell to 11.8 million acres in 1939.

The arguments about State subsidies and guaranteed prices were slow to be resolved; indeed they remained unresolved by 1939 in spite of the fact that during the 1930s pressure for increased intervention in social and economic affairs had been stepped up. The new National Government pledged a comprehensive system of protection, but an emergent policy of Imperial preference, whereby Empire goods were exempt from paying import duties, was scarcely effective. In fact steps to alleviate agricultural distress were *ad hoc* and limited: for example the Wheat Act of 1932 promoted some recovery of the area under wheat, and Marketing Boards were set up (seventeen by 1939), of which the best known was the Milk Marketing Board, established in 1933 to regulate a substantial increase in liquid milk production. But a strong case remained for relying on other methods such as self-help, improvement of credit facilities, measures to promote farming efficiency, and a readiness to accept a certain drift from the land. The fact was that the case for subsidies was not overwhelming; after all the urban population gained enormously through cheap food, and the country could afford the import costs.

The general scene then was of painful contraction and a widespread reduction in farming expenditure. Commentators reported a lowering of cultivation standard, evidence of impaired drainage, overgrown hedgerows and countless buildings in disrepair. Agriculture was in depression, farm wages low and housing conditions unsatisfactory. Moreover, the practice of farming was in poor shape for a capitalized industry. Orwin, writing in

1945, could observe that 'taking all types of farming together, it is a fair generalisation to say that on nearly half the land of Britain (47 per cent) the standard of efficiency in food production is condemned to be a long way below the maximum attainable' (p. 24). Regarding local amenities, almost half the parishes of England and Wales lacked a sewerage system, and a third had no piped water; electricity reached less than one in twelve farms in 1938 (Scott, 1942).

Changes, however, were afoot. Milk was being actively promoted as a health food; there were increases in production in fresh fruit and vegetables; and pig and poultry rearing was intensified, as it made a transition from cottage industry to factory farming. Tractors were more plentiful and the beginning of the end was in sight for horse husbandry (though in 1939 there were still nearly 650,000 horses employed in British agriculture, doing about two-thirds of the total work (Creasey and Ward, 1984)). Further, a positive initiative had been taken to rectify the serious shortage of timber, revealed during the Great War. The Forestry Commission was set up in 1919, charged with afforesting 1.75 million acres. Progress was slow, however, and its planting programme was incomplete by 1939.

The effect of all this on the rural population was easy to see. A quarter of a million men left the land between the wars as the agricultural workforce fell by about one-third. But while many observers urged the need for positive measures to halt this exodus and to regenerate rural life, the fact was that the number of workers involved was considerably less than the extent of urban unemployed in the basic industries. Indeed, for many years the number of persons registered as out of work regularly exceeded the total number of persons engaged in agriculture. Hence, while the 1921 Census recorded only 7.2 per cent of the national workforce as engaged in agriculture (a very low figure compared with 38 per cent for France and 30 per cent for Germany), this particular rural question did not receive the political attention it otherwise deserved. As will be noted later, there were other countryside concerns which were accorded greater popular weight.

But it was still a fact that because many of the older rural industries had failed to survive earlier periods of competition from new technologies and different scales of production, the residual economic base of rural England and Wales was particularly fragile. A Rural Industries Intelligence Bureau, set up in 1921 under the direction of the Ministry of Agriculture but funded through the Development Commission, could give only limited aid. (In 1923 the 'Intelligence' part of the title was dropped, after which it was known as the RIB or simply 'The Bureau'.) It proved extremely difficult to revive traditional craft industries, and apart from a steady demand for quarrying to supply the construction and road building industries, the expansion of mining (iron ore in Northamptonshire, coal in Kent), and the spill-over effect of new factories particularly on the London fringe, it was a picture of sustained economic gloom. There was nothing to compare with the potential of hydro-electric power developments in Scotland during the 1930s.

Alongside these difficulties, the early 1920s recorded remarkable changes in land ownership. As seen in chapter 3, the growing burden of death- and estate-duties encouraged the sale of large areas of rural land. It would be difficult to exaggerate the importance of the changes that resulted: between 1918 and 1922 around 25 per cent of the agricultural land of England came into the possession of those who farmed it. This was a sale of land unprecedented since the dissolution of the monasteries in the sixteenth century (Newby, 1987). The proportion of farmland in owner occupation already stood at 20 per cent in 1921 (having risen from 11 per cent in 1913); by 1927 it was 36 per cent. The alarming fall in agricultural prices in the later 1920s and early 1930s induced heavily mortgaged farmers to further sell their land. The result was a marked shift in social and economic power away from the aristocracy and the gentry to the farmers. Some of the new buyers were speculators or people who had done well out of the war; some were Scots looking for good farms on cheap land, but most were the farmers themselves who became small landowners, masters of the parishes where they lived (Howkins, 1991).

These tenure changes received little attention for some time. Meanwhile the old agenda of agricultural reform and revivification of rural life held sway in political circles. The Liberal Land Enquiry Committee had surveyed the rural land problem in the period 1909–12, publishing its findings in 1913. After the war, a successor, the Liberal Land Committee, published its rural report in 1925, *The Land and the Nation*.

Their starting point was a development of an earlier conviction.

> The prosperity of the town depends on that of the country; that of the nation on both . . . A nation relying on its manufacturing industries is unstable, particularly when unemployment in industries threatens to become permanent and when our command of the world markets will not return. We have more people unemployed than we have in agriculture . . . We have a smaller proportion of our people living on the land than any country has ever had. We depend for our food on others more than any country depended before. We have a landless peasantry such as nowhere else has existed, and we offer them too little in the way of independent life and practically nothing in the way of chances for advancement. (p. 3)

The Committee's argument was that 'agriculture, if it were adequately capitalized and vigorously led, could win a higher production from the land and maintain on the land a larger population' (p. 195). To achieve this, it suggested that the State should resume possession of all land used 'for the production of foodstuffs, timber or other natural products' and that farm holdings should be transferred to any person 'competent to use it to the advantage of the community as a whole', the land ownership to be held in 'Cultivating Tenure' (p. 299). An accompanying superstructure of reform measures would include provision for agricultural credit; research and education; marketing and transport; rural industries, forestry and social life; and survey and reclamation.

This particular set of proposals advanced no further. But it is worth recalling them because they link with earlier socio-political reforms and ideas for dealing with the rural problem of agriculture and land and provide a bridge to other solutions adopted in later years.

COUNTRYSIDE RECREATION

A largely new set of issues now came to the fore in a series of campaigns for freedom to engage in active leisure pursuits, as represented by hikers and ramblers. In chapter 2 it was noted that enclosure had resulted in the loss of many rural footpaths, and that the Commons Preservation Society had been instrumental in saving access to a number of commons in and around London. About the turn of the century the question of access to hill land and moorland in the north of England highlighted a new concern. Towards the end of the nineteenth century public access was restricted from those areas chosen to become game reserves, shooting having become a major pastime of the aristocracy and landed gentry. Hence in Scotland deer forests, and in England grouse moors were closed to local walkers by gamekeepers.

The leisure pursuit of walking and rambling was first taken up by professional and generally middle class people. But it quickly spread to the working class through a loose organization of rambling clubs, and during the 1920s increasing support came from the northern industrial towns – particularly those adjacent to the heather-clad Pennines, an area of prime grouse shooting. Between 1926 and 1932 rambling peaked in enthusiasm and militant trespass in the southern Pennines marked the clash of interest between on the one hand the landowner and sporting tenant, and local authorities and water works authorities, who were equally unsympathetic, and on the other hand the working class rambling clubs. Federations of rambling organizations were founded in Liverpool (1922), Manchester (1926) and Sheffield (1926) and others followed later.

The flashpoint of conflict was access to the Peak District where there were only twelve footpaths which exceeded two miles in length, and particularly to the Kinderscout Plateau where there was no public footpath and access was rigorously prohibited by its owner. The nesting period (April to June) and the shooting season (12 August to 10 December) provided the occasion for clashes between gamekeepers and trespassers.

But the Rambling Federations mounted a vigorous campaign for access, setting out a charter for ramblers. Rallies were held, a favourite location being the Winnats Pass near Castleton, and weekend trains brought in the urban ramblers – some unemployed and hence with special travel facilities. Militancy led to organized mass trespasses, of which that on Easter Day, 24 April 1932 in the region of Kinderscout by a crowd of 600 people led by one Benny Rothman, organized by the British Workers' Sports Federation (BSWF), a body set up and led by the Young Communist League, was the most celebrated (Rothman, 1982). There were brief scuffles with game-keepers half way up the face of Kinderscout, a celebratory meeting was held at the top, and on the group's return to Hayfield policemen made six

Outdoor recreation between the wars. Walkers at the top of Styhead Pass in the Lake District, 1922. (By courtesy of the Rural History Centre, University of Reading)

arrests. Those charged appeared in court the next day and, after remand, again on 11 May. There was one charge of causing grievous bodily harm to a keeper, otherwise the charges were those of unlawful assembly. At Derby Assizes on 21–22 July the charges related to riotous assembly and assault. One of the six was discharged; for the rest the sentences ranged from two to six months.

Rothman's account had been challenged by Stephenson (1989), a pioneer in the agitation for the legal right of access to the countryside, who became involved in the organized rambling movement. Tom Stephenson's autobiographical record makes it clear that various Ramblers' Federations stood aloof from the activities of the BSWF, a body which was not known to have had any previous interest in the access problem, and which failed to play any part in the subsequent access campaign. Moreover, 'the truth is that there never was a mass trespass. No one reached the summit of Kinderscout and the so-called victory meeting was held on a public path at Ashop Head' (p. 153). Rothman's party, according to Stephenson, were two miles to the north west of Kinderscout and about 400 feet lower than the highest point. In any event the episode caused great controversy at the time, a belief (current but challenged) being that the organized disruption by the BSWF had hindered the negotiations then under way to secure access to mountain and moorland.

Rambling enthusiasm subsequently waned and the radicalism of the Federations declined. But not before a major social force had emerged. C.E.M. Joad (1946), philosopher and ramblers' friend, wrote of the

> living witnesses of this revolution at the Central Station at Manchester on a Sunday morning, complete with rucksacks, shorts and hobnailed boots, waiting for the early trains to Edale, Hope and the Derbyshire moors . . . In our day hiking has replaced beer as the shortest cut out of Manchester, as turning their backs upon the cities which their fathers made, armies of young people make sorties at any and every opportunity into the countryside. (p. 17)

The Ramblers' Association was formed in 1935 as a federation of a host of local rambling clubs referred to earlier, and attention turned to negotiation with the landowners and the promise of an Access to Mountains Bill, introduced to Parliament in November 1938. This was the eighteenth Private Members' Bill put forward to gain unlimited access, as of right, to the mountains of Scotland or of Britain. Only two had ever reached a second reading, but because of particular circumstances of the time, Arthur Creech-Jones' Bill succeeded in complicated horse trading between the parties involved (Cherry, 1975). But the 1939 Act was quite without effect: not one Access Agreement was made under its provisions. The country was soon at war and national circumstances changed again.

Rambling of course was only part of the inter-war pre-occupation with mass, outdoor leisure pursuits. The Cooperative Holidays Association, established in 1891 by T.A. Leonard, was followed by the Holiday Fellowship in 1913, both organizations set up to provide cheap holiday accommodation. Numbers attending holiday centres increased between the wars. Even more dramatic growth was reflected in the growth of the Youth Hostels Association, founded in 1930.

But perhaps the most spectacular development in holiday provision was in organized camping. The Victorian and Edwardian seaside resort was already fixed in popular pleasures, the manifold variety of pleasure buildings quite remarkable in extent as Pearson (1991) demonstrates, and for some time in the early years of the twentieth century the popular resorts were feeling their way to a freer approach to the enjoyment of leisure time. Walton (1983) writes that 'mainstream middle-class and working-class culture had become reconciled in mutual tolerance at those resorts where the classes regularly mingled . . . the content and control of working-class leisure had become a matter of indifference rather than a cockpit of controversy' (p. 215). The time was propitious for a new initiative, and it came with organized holiday camping.

We have already referred to the early twentieth-century interest in sleeping under canvas; this was maintained by the uniformed organizations and by the burgeoning membership of the Camping Club of Great Britain. The holiday camp entrepreneur, heralded by Joseph Cunningham in the Isle of Man, came into his own in the 1930s (Ward and Hardy, 1986). The 'pioneer' camps, including school camps and workers' camps, were gradually replaced by the new commercial camps. Harry Warner began his holiday camp empire with a site at Hayling Island in 1931; Billy Butlin opened his Luxury Holiday Camp at Skegness in 1936, followed by a camp at Clacton in 1938 and at Filey, scheduled to open in 1940. A social revolution of a kind had taken place; the seaside landlady was giving way to another form of holiday industry as, by 1939, perhaps a million and a half people spent their holidays under canvas and in camps. The Holidays with Pay Act, 1938 considerably increased the number of workers benefiting from paid holidays (affecting 18.5 million employed workers, nearly 11 million of whom were receiving holiday pay for the first time), and this addition to leisure time impacted significantly both on the countryside and on the coast.

It was estimated that an aggregate of 13 million people may have taken over one week's holiday away from home in 1937, while a survey in 1939 suggested that about 45 per cent of holidays were spent by the sea (Sheail, 1976*a*). Between the wars the coast and the rural hinterland was under increasing pressure from development, and very often of an objectionable kind. As extensions of the former plotlands (Hardy and Ward, 1984), huts, caravans, old railway carriages and bus bodies ribboned many coastal stretches, and bungalows and shacks lined much of south east England. Peacehaven was developed by an entrepreneur, Charles Neville, in ploughed fields on the cliff tops between Brighton and Newhaven; by 1935 650 dwellings had been erected at very low density in a massive grid. Shoreham Beach was a singular mess of railway carriages and timber bungalows. In Essex Jaywick Sands was developed by another enterprising speculator, Frank Stedman, out of several hundred acres of reclaimed marshland, in spite of the opposition of nearby Clacton. Canvey Island as another creation of an individual entrepreneur, Frederick Hester; across the Thames Estuary, Minster-on-Sea on the Isle of Sheppey, developed by Frederick Ramuz was less successful. Elsewhere, over many stretches, rural and coastal disfigurement came from advertisements, the haphazard erection of unsightly buildings, cafés, wayside stalls and sporadic camping sites. As a portent for an additional problem, the first factory production of the trailer caravan took place in 1922.

National Parks

A combination of factors conspired to focus on one further development in this inter-war period: the establishment of National Parks. These included: a combination of a general interest in outdoor recreation by increasing numbers of people across the social classes; a health campaign and attendant promotion of the virtues of an open air life; new opportunities to enjoy the countryside and coast; and a determination to gain access to private land for the purposes of rambling. In the context of a growing unease by government at the rural disfigurements and at the inability of local authorities (except under special provisions) to control unacceptable development, the Prime Minister (J. Ramsay MacDonald) appointed a Committee of Inquiry in 1929 'to consider and report if it is desirable and feasible to establish one or more National Parks in Great Britain with a view to the preservation of natural characteristics including flora and fauna, and to the improvement of recreational facilities for the people' (Cherry, 1975). The chairman was Christopher Addison, a former Minister of Health and currently Parliamentary Secretary to the Minister of Agriculture.

The Committee reported in April 1931 (*Report of the National Park Committee*). While National Parks in the North American sense were not practicable in Britain, the Committee did strike a positive note. They found in favour of a system of 'National Reserves' and 'Nature Sanctuaries' in order to safeguard areas of exceptional national interest against disorderly development and speculation; also to improve the means of access for those on foot to areas of natural beauty, and to promote measures for the protection of flora and fauna. No government action ensued, but, as often in these cases, it proved more significant that a very influential lobby had been encouraged.

In 1936 the Councils for the Preservation of Rural England and Rural Wales set up a Standing Committee on National Parks; with a membership including representatives of many amenity organizations, its chairman was the lawyer Sir Norman Birkett. The Standing Committee proceeded vigorously to espouse the Addison Report. Their pamphlet, *The Case for National Parks in Great Britain*, published in 1938 was typical of its forceful views expressed in a way to secure popular support. In the foreword Professor G.M. Trevelyan wrote of the need 'to preserve for the nation walking grounds and regions where young and old can enjoy the sight of unspoiled nature . . . Without vision the people perish and without sight of the beauty of nature the spiritual power of the British people will be atrophied.'

At this point the campaign for National Parks was overtaken by a resumption of the agitation for 'access to mountains' which culminated in the surprising success of a Private Member's Bill introduced by Arthur Creech-Jones, MP for Shipley. This secured a measure of agreement in Parliament; it went through the House and received the Royal Assent in July 1939. The concessions wrought from the landed and sporting interests were as much as could be achieved, but the problems relating to footpaths, rights of way, rambling access, and in a broader context National Parks, were looked at again quite shortly, though in very different circumstances. In the meantime, the countryside recreation question between the wars had provided a new set of concerns for rural England and Wales and posed dilemmas in public policy for a government slow to meet demands for an extension to State involvement in social and enviromental affairs.

BUILDING IN THE COUNTRYSIDE

Four million dwellings were built in Britain between the wars, equivalent to about one-third of the total housing stock by 1939. Evidence given in the Scott Report (1942) suggests that in the period 1927–39 around 60,000 acres a year were taken for building and other constructional development. This was an average figure; in the late 1920s and early 1930s it was lower, but in the later 1930s considerably higher due to the demands of the War Office and the Air Ministry. One of the major features of the countryside inter-war, therefore, was the significant absorption of rural land for urban purposes. This was another perspective of an evolving rural agenda, for while the territorial spread of cities was not new, the new scale of change was exceptional.

A number of factors accounted for this development (Cherry, 1988). Opinion had long held that the Victorian city was insanitary, overcrowded and unhealthy, and public regulation of housing aimed to increase space standards and reduce densities. The late Victorian suburb was a reflection of this. But a significant departure in residential building norms took place around the turn of the century with an enthusiasm for cottage architecture design and low density, informal layouts. The garden suburb was born and after 1919 this form of housing estate layout became ubiquitous.

Another factor was that the State itself took on new powers to provide public housing. Significantly, the form of that housing followed the popularity accorded to low density. As noted in chapter 3, Raymond Unwin's contribution to the seminal Tudor Walters Committee Report (1918), relating to the provision of dwellings for the working classes, was significant in this respect. The government's *Housing Manual* (Local Government Board, 1919) adopted important recommendations as to density (12 to the acre in urban areas and eight in rural areas), generous space standards and overall layout. The consequences made their mark over the next two decades. One in four of the dwellings built in the country between the wars was a council house, and council estates constituted a sizeable element of the demand for rural land both on the outskirts of cities and in smaller sites elsewhere.

The main building surge, however, came from the private sector, with a new-found preference for owner occupation, but equally disposed towards low density housing. The cost of building land was low, given the readiness of farmers to sell, and there was an increasing capacity of a salaried, expanding middle class to contemplate purchase in preference to renting property. The Building Societies proved able institutions in lending money, and ownership became more and more a prudent venture. As a consequence, speculative building boomed and the inter-war years threw up a new, distinctive suburban environment comprising largely three bedroomed, semi-detached dwellings, with small front gardens, larger rear ones, drives, hedges and garden gates, symbolized in 'Dunroamin' ' (Oliver, Davis and Bentley, 1981). In these circumstances, the speculative builder boomed: New Ideal Homesteads, Laing and Costain, Taylor Woodrow, Wates and Wimpey, to name some of the larger ones (and most were in fact very small).

Another factor related to transport. Around London, improved and more extensive links permitted living further afield from a work place than ever before. The extraordinary expansion of London Transport, particularly its underground lines, was a feature of metropolitan life in the 1920s and 1930s. Lines reached out to all points of the compass, though particularly to the west, north and east, as new rural areas were opened up for development. The potential had already been demonstrated by the speculative activity of the Metropolitan Railway Company and its marketing flair in north-west London (Green, 1987). 'Metro-land' was the

Ribbon development into the countryside – urban growth to the west of Sheffield in the mid-1930s. This picture was taken by the local branch of the CPRE and the original caption talked of 'box-like monotonously spaced semi-detached villas which spoil a beautiful skyline'. (By courtesy of the Rural History Centre, University of Reading)

creation of the Company's publicity department which adopted the name in 1915 as a catchy shorthand for that sector of the metropolis. It was also the title of a guide book issued annually by the Company between 1915 and 1932, the heart of which was the house-seekers section.

The Company was absorbed by the London Passenger Transport Board in 1933, but not before its annual publication had been instrumental in marketing many new houses in London's north-western rural fringe. The Metropolitan Railway had been directly involved in house development for some years, utilizing surplus land acquired before the railway was built, as at the Willesden Park Estate in the late nineteenth century and in the early 1900s at Pinner and Wembley Park. But in 1919 a new property company was created to manage and develop the railway's estates and over the next 13 years a series of new residential areas was developed progressively at Neasden, Wembley Park, Northwick Park, Eastcote, Rayners Lane, Ruislip, Hillingdon, Pinner, Rickmansworth and Amersham. London's periphery burst into the surrounding countryside, absorbing earlier settlements in tentacles of building development, threaded by the transportation corridors of road and rail.

For a number of reasons, therefore, rural land was the subject of urban pressure for building purposes. Local authorities built their suburban housing estates, some over extensive tracts as at Kingstanding north of Birmingham, Speke at Liverpool and Walker east of Newcastle; Manchester developed the Tatton Estate for its suburb at Wythenshawe; and the LCC had the biggest country estates of all, notably at Becontree in Essex. The private sector built massively, particularly in the more prosperous areas as in south Birmingham (Cherry, 1994), but above all around London (Jackson, 1973). The population of Middlesex increased by 1.64 million between 1921 and 1931, the largest rise recorded for any county in that decade. The district of Harrow absorbed nearly 135,000 people between 1921 and 1938, the largest influx of any local authority in Greater London. By the end of the 1930s all but the north and north-western parts of the county had been built over, the flat clay lands ravaged by the building companies which exploited the accessibility afforded by the railway and (increasingly) the motor car. The countryside of north-west Surrey was built over and so too was the coastal belt of Essex. Further afield Hertfordshire showed growth around Barnet, and Buckinghamshire around Slough. Finally, in the late 1930s demand for aerodromes introduced a new factor in land take-up.

The *Report of the Royal Commission on the Distribution of the Industrial Population* (1940), the Barlow Report, noted the urban impact on rural England:

> The belt of the traditional 'corn counties' has been seriously reduced by the growth not only of Greater London but also of the much wider residential zone around it. In South Essex, Middlesex and North Kent the belt included market gardening, fruit orchard and potato-growing areas now rapidly vanishing; and areas of exceptional importance devoted to dairying and potato-growing in close proximity to industrial centres in Lancashire, Cheshire and the Midlands are endangered by the zoning for industry and residences of the belts forming the peripheries of the great conurbations. The amount of first-class land in Great Britain suitable for market gardening is limited; it has been estimated that the amount of land at present intensively used is only about 2 per cent of the whole and that only about 5 per cent could be so used. Yet much of it runs severe risk of being absorbed by industrial and housing development, which since 1900 has been so rapid that it is stated to have covered with bricks and mortar an area equal in size to the counties of Buckingham and Bedford combined. (para 36)

The Royal Commission elsewhere noted the centrifugal consequences of population dispersal from London, observing that in recent years some county towns near London, such as Chelmsford, Bedford, Luton and Aylesbury, had shown a growth of insured population faster than London's outer fringe itself. The vigour of this outer ripple of metropolitan growth presaged even greater pressure on rural land. In a sense, however, this was precisely what one lobby would wish to see: since World War I the garden city movement had strongly espoused the wholesale decongestion of London and the big cities in favour of a dispersed population strategy.

The changing rural scene – the invasion of the motor car. This view of the village of Husband's Bosworth, Leicestershire in the mid-1930s shows the effects of the motor car upon the countryside. The RAC man directs an outdated form of transport. Note the rash of advertising signs which were a particular concern of the early CPRE groups. (By courtesy of the Rural History Centre, University of Reading)

Hardy (1991) remarks that the Garden Cities and Town Planning Association's Memorandum to the Royal Commission, prepared by F.J. Osborn, was 'a closely-argued case, probably the most important document for the movement since Howard's original book on garden cities' (p. 199).

After Letchworth, only one further Garden City was built – at Welwyn, commencing in 1920 – but the principle of a national programme for planned satellite towns was increasingly advocated by the G.C.T.P.A. Beevers (1988) argues that in fact the Association was 'taken over' by a small caucus devoted to this cause. A National Garden Cities Committee working within the Association won control and gave an uncompromising stamp to an unremitting campaign for garden cities, satellites and new towns. From the mid-1930s the Association, now led by Gilbert McAllister, as Secretary and Editor, and Frederic Osborn as Honorary Secretary, adopted a wider goal of national planning as the strategic vehicle for population redistribution. This spoke the language of the Hundred New Towns Association, founded by the architect A. Trystan Edwards in 1934 which advocated, as the name suggests, one hundred new towns to relieve pressure on the metropolis; in a migration of five million people, no less than 40 of the new towns were to be south of a line from the Wash to the Severn.

Though nothing immediate came of this, in 1939 all the indications were that rural England (in particular) had experienced an alarming two decades from the point of view of building development, and that the future offered neither significant modification nor reversal of the trends. Uncertainty prevailed; there was a widespread acceptance of the advantages to be derived from decentralization, but little prospect of a purposive strategy to achieve it. In the meantime the countryside was at risk from the spread of building development only feebly controlled.

One particular risk was largely hidden from view: the pollution of underground water. Between the wars important political initiatives were undertaken by a number of urban local councils with regard to sewage disposal in rural fringe areas. Brighton Corporation, for example, came to regard speculative building development in its hinterland on the South Downs as constituting a grave threat to the purity of its water supplies. Through a Parliamentary Bill (1924) it sought unprecedented powers to acquire land and regulate the provision of sanitation over a wide area (Sheail, 1992*a*). Rather surprisingly the Bill was successful and the Brighton Corporation Water Act, 1924 conferred powers to acquire compulsorily over one thousand acres of downland in the vicinity of Patcham Pumping Station and to impose bylaws for the protection of lands surrounding that and other stations in order to prevent the percolation of sewage into the underlying chalk.

Other political initiatives were connected to improvements in sanitary infrastructure over extensive suburban areas. The sewage farm and forms of biological filtration had evolved from agricultural practices in the nineteenth century, but the activated sludge process was the first to be created scientifically for sewage treatment. This new technology made it possible to envisage schemes that drained entire catchment areas into one disposal works. The Birmingham, Tame and Rea District Drainage Board, established in 1879, provided the model for coordinated sewage disposal over a large area (in this case nine separate authorities within three counties). Between the wars the unparalleled suburban surge highlighted the need to secure optimal methods for sewerage and waste disposal. Sheail (1993*a*) traces the steps taken in North Cheshire, consequent upon the development of Wythenshawe, the pioneering measures taken by Middlesex County Council (the West Middlesex Drainage Scheme) and the further work in Hertfordshire for the drainage of the Greater London area. These and other developments presaged the emergence of specialist authorities for water supply, sewage treatment and disposal, which the post-war years would address.

PRESSURE FOR CONSERVATION AND PROTECTION

The countryside was clearly experiencing profound change: the agricultural base was in decline and seemed incapable of recovery, there were pressures to accommodate escalating forms of mass recreation, and visually, both the impact and the sheer scale of building development aroused widespread disquiet. There were no immediate, obvious remedies, the State still hesitant to extend and sharpen its powers of regulation, let alone determine to enter the largely uncharted waters of strategic land-use planning. For a while, therefore, protest had to pursue its course and the running was made by a loosely related network of nature conservation bodies, environmental groups and countryside lobbies united in their dismay at the pace of unwanted change in the environment and the loss of a treasured past. The inter-war period therefore saw the consolidation of a number of voluntary amenity bodies, and indeed the emergence of new ones; it also threw up some important key writers who articulated and helped to shape popular opinion. This formative period was crucial as a preparation for the greater powers that the State would soon be obliged to take up.

A number of important amenity bodies were already active, including the Commons, Open Spaces and Footpaths Preservation Society, the National Trust and the Society for Checking the Abuses of Public Advertising. A new influential body was founded in 1926: the Council for the Preservation of Rural England (CPRE), the CPR Wales following shortly afterwards. It was the joint initiative of Professor Patrick Abercrombie in his year as President of the Town Planning Institute, Guy Dawber, President of the Royal Institute of British Architects, and the Earl of Crawford and Balcarres, Chairman of the Fine Arts Commission (Cherry, 1974). It succeeded in becoming a focal body for a disparate range of special interest groups, and, as already observed, its authority was such that it was able to set up the influential Standing Committee on National Parks, which soon assumed the lead in that area. With twenty-one bodies as constituent members of the Council, and a larger number affiliated, there was now a new liaison between all those concerned with countryside preservation.

One example of the CPRE's rapidly acquired status, and an illustration of its coordinating role in the absence of intervention by central government or local authorities, concerned the development of the Breckland region of East Anglia between the wars. The Forestry Commission acquired its first property in this grass heathland in 1921; plans envisaged the large-scale planting of Corsican and Scots Pine for pitwood and pulp (Sheail, 1993*b*). Concern at the extensive loss of natural habitat led to the Forestry Commission and the CPRE forming a Joint

Informal Committee in 1935; they subsequently approved a plan prepared by the landscape architect, Geoffrey Jellicoe, which envisaged the forested areas of Breckland interspersed with large open spaces formed by the famous heaths, and interconnected by a 30-mile long footpath around Thetford.

Concern over the increasing despoilation of the environment came from a variety of different sources. The architect Clough Williams-Ellis (1928) in 'an angry book written by an angry young man', as it was described in a much later edition when the author was in his ninety-second year, tilted at the follies and abuses of the late 1920s. He likened the urban spread to the tentacles of an octopus and deplored 'the spate of mean building all over the country that is shrivelling up the old England' (p. 15). He observed drily 'In the late war we were invited to fight to preserve England. We believed, we fought. It may be well to preserve England, but better to have an England worth preserving. We saved our country that we might ourselves destroy it' (p. 20). He argued that 'We are modifying both town and country, removing the worst reproaches from the one and much of its essential charm from the other. We plant trees in the town and bungalows in the country, thus averaging England out into a dull uneventfulness whereby one place becomes much the same as any other.'

This sense of destroying the countryside through a failure in town building was taken up by another angry young man, the planner Thomas Sharp in two remarkable books written in the 1930s. A 'man who dared to be different' (Cherry, 1983), he in particular would leave a legacy to the town planning movement in his prescription for the sharper differentiation between town and country. He offered a description that was widely shared. In *Town and Countryside* (1932) he bemoaned that

> From dreary towns the broad, mechanical, noisy, main roads run out between ribbons of tawdry houses, disorderly refreshment shacks and vile, untidy garages. The old trees and hedgerows that bordered them a few years ago have given place to concrete posts and avenues of telegraph poles, to hoardings and enamel advertisement signs. Over great areas there is no longer any country bordering the main roads: there is only a negative, semi-suburbia . . . There seems no valley or hillside that is not despoiled by some shameless blot. Gigantic pylons stride over the counties, like the mechanical Martians of Mr H.G. Wells's *War of the Worlds*. Commons, moors, woods, hedgerows are littered with refuse. Everywhere one expects ugliness, incongruity, and disorder. (p. 4)

His analysis was a more personal one, arguing that we had the worst of all worlds – the degradation of the town and the destruction of the countryside.

> Two diametrically opposed, dramatically contrasting, inevitable types of beauty are being displaced by one drab, revolting neutrality. Rural influences

> neutralise the town. Urban influences neutralise the country. In a few years all will be neutrality. The strong, masculine virility of the town; the softer beauty, the richness, the fruitfulness of that mother of men, the countryside, will be debased into one sterile, hermaphroditic beastliness. (p. 11)

His prescription followed logically, to preserve a distinction between town and country; the countryside would be saved by preserving urban traditions of scale and compactness and rejecting low density, semi-detachment.

In *English Panorama* (1936) he followed the same vein. Towns were social units to live in, not to spread out from, and our failure to recognize this was destroying the countryside and its landscape of composed quality – its intimacy, sense of order and design, rhythm, and quietness of age and custom.

Williams-Ellis and Sharp were two professionals, no doubt with an axe to grind. Certainly today it would be easy to be dismissive of their views as misguidedly romantic of a mythical rural past and so out of touch with the trends of the day that their advocacy for the future was unreliable. But that is not the point. In the 1930s they were articulating widely held prejudices, and in the absence of other certainties their angry rejection of the forces of modern civilization and their destructive environmental effects struck responsive chords.

Another radical, the philosopher C.E.M. Joad, lover of the English countryside, also railed against the awfulness of rural despoilation. In a memorable passage (1946) he described a pre-war scene in the Chilterns. In a long winding valley running up from Marlow he found

> a muddy road running through an avenue of shacks, caravans, villas, bungalows, mock castles, pigsties, disused railway carriages and derelict buses . . . Each dingy little abode in this rural setting was distinguished by some dreadful appellation, as, for example, Eretiz, The Nest, The Splendide, Kosy-Kot, Mon Abri, Linga-Longa, or U-an-I, as if to throw into even grimmer relief the dreariness and the shabbiness of the dwellings which facetiousness sought to enhance or pretentiousness to dignify.

He continued walking, only to find himself 'in a world of rural factories where ladies' corsets and electric light bulbs competed for the privilege of eating up what was once first-class agricultural land' (p. 37).

The passionate rejection of urban mores and their influences on a treasured countryside was obvious. If we are to understand subsequent war-time and post-war developments in rural planning, we do well to appreciate that depth of concern so broadly shared in the 1930s. The reaction was formidable. The launching of the magazine *The Countryman* in 1927, and the Batsford books which popularized the beauties of the countryside, showed the extent of a receptive readership. One popular writer was H.J. Massingham who inherited the mantle of earlier literary guardians of rural traditions; writing prolifically he progressed from observer to committed polemicist, intuitively resistant to urban influences on the countryside (Abelson, 1988).

Meanwhile the conservation ethic was being promoted from other quarters. The geographer Vaughan Cornish (1937) eulogized the heritage of scenery, drawing attention to this country's unique qualities. The coastal scenery was of course particularly valued, and in the absence of any government action on scenic despoilation, the CPRE, the Commons, Open Spaces and Footpaths Society, and the National Trust jointly set up a Coastal Preservation Committee in 1937 to highlight the unsatisfactory situation which then obtained.

A more important source of influence (certainly with regard to nature conservation) was that of the natural scientists and regiments of amateur collectors and observers who soon formed their own societies. The Society for the Protection of Birds, formed in 1889, was granted a Royal Charter in 1904. The Society for the Promotion of Nature Reserves dated from 1912. These bodies and the various County Naturalists' Trusts were increasingly active in the 1930s (Sheail, 1976*a*) and complemented a very vigorous lobby for conservation and the preservation of rural amenities.

Another concern was water pollution. The country's fresh water fisheries had undergone decline since the last decades of the nineteenth century as polluted rivers, particularly in the urban and industrial heartlands, showed falling fish stocks. Mounting concern led to the Ministry of Agriculture and Fisheries appointing a Standing Committee on River Pollution in 1921. Chaired initially by the Minister and subsequently by the Fisheries secretary, it was composed of representatives of the salmon, trout and coarse fishing interests, and of the Federation of British Industries. The Salmon and Freshwater Fisheries Act, 1923 introduced new powers to tackle pollution, but more scientific work was needed and a Water Pollution Research Board was appointed in 1927, under the aegis of the Department of Scientific and Industrial Research – the forerunner of the Science (and Engineering) Research Council.

The inter-war years were an important period in the development of an institutional and research response to pollution issues. By 1939 the Board's work had established a sound reputation, with detailed surveys of the River Tees and the Mersey estuary. There were also investigations of two new problems relating to the treatment of beet-sugar and milk effluents (Sheail, 1993*c*). A large-scale beet-sugar processing industry developed in eastern England in the 1920s, the amount of beet-sugar manufactured rising from 7,000 tons in 1922–23 to over 500,000 tons in 1938–39. With a continuous washing operation in the manufacturing process, waste waters were discharged into streams or rivers. Rapid decomposition of organic matter reduced the concentration of dissolved oxygen to a level low enough to destroy fish and aquatic life; in the 1920s the River Lark below the Bury St Edmunds factory became almost completely deoxygenated. The treatment of milk effluent was also a problem with the increase in size of depots: before 1914 few handled more than 5,000 gallons a day, but this throughput soon reached 50,000 gallons. Between a third and two-fifths of the 900 million gallons processed in the depots in the mid-1930s was used for manufacturing purposes, giving rise to the problem of waste waters. These examples of water pollution from two rural manufacturing sources showed how the national problem was becoming not only more geographically extensive, but more complex in its ecological impact. The conservation debate of the post-war years was in the making.

Another concern was land despoilation through mining operations in the countryside – not the legacy of waste from deep-mined coal as in east Durham, Yorkshire or the east Midlands, but the new scale of open-cast iron ore mining, particularly in Northamptonshire. Mechanical excavators had been used as early as the 1890s to strip the overlying burden containing oolitic limestone. But a dramatic escalation of the operation came during the 1920s when the countryside around Corby was targeted (Sheail, 1983). In 1932 a Scottish steel-making firm, Stewarts and Lloyds, established a new works there, based on the local ores. Previously the ironstone field had been exploited where the ore occurred in outcrops, or within 20 feet of the surface, in which case restoration to agricultural use presented little difficulty. But with a new generation of power shovels 60 feet of overburden could be removed over extensive areas; by 1936 1,600 acres had already been transformed into hill and dale and there were fears of many square miles of Northamptonshire being transformed and rendered derelict, with little prospect of restoration. Once again we see that the potential scale of environmental problems in the countryside was mounting in such a way as to cause increasing alarm. In many cases no clear solutions were yet in sight.

REGULATION AND PLANNING

During the years between the wars Britain took tentative steps to establish a national land use planning system. Action was slow and indecisive; powers were weak, local authority effectiveness heavily circumscribed, and there was a political inability to grapple with the financial questions of compensation and betterment. But by 1939 there was a recognition that many of the problems of the countryside (and of urban areas too, of course) required a stronger planning system for their resolution; the trials and tribulations of two decades provided valuable experience for the time when a comprehensive sharpening of legislation proved possible. Meanwhile professionals working in planning matters had improved their own capabilities and reputation, and there was a small cadre of academics and consultants, of whom Patrick Abercrombie had the greatest reputation. We should regard the inter-war years as a formative period as much as simply a wasted opportunity created through bumbledom and political chicanery. Everything that came after depended on these years of experiment, and there were positive lessons as well as negative ones from which to learn.

Town planning initially was all about preparing layout schemes, whereby small areas of land in the vicinity of towns could be developed in such a way as to secure proper sanitary conditions, amenity and convenience. The 1909 Act had aimed to provide no more than a measure of regulation for the Edwardian suburb. After the First World War the essence of this procedure was maintained, though simplifications were introduced. An immediate national problem was to engage in an extensive house building programme, and the preparation of town planning schemes was judged necessary to ensure the proper development of urban extensions. Statutory town planning was still concerned with the siting and layout of new housing areas – existing built-up areas and the open countryside were excluded.

The Housing and Town Planning Act, 1919 required the formal resolution of a local authority to prepare a scheme in respect of an area of land; after 1 January 1923 the preparation of schemes was compulsory on all Urban Councils with a population of more than 20,000. Thus it was as one example, that more than 3,000 acres of farmland in east Birmingham was 'town planned' in that part of the city as part of the Birmingham City, North Yardley and Stechford Town Planning Scheme, approved in 1921, and its various residential estates progressively developed (Cherry, 1994). But local authorities were slow to make their proposals and by 1928 there were still 98 out of 262 (with a population of more than 20,000) which had failed to submit any proposals either in the form of resolutions, statements or schemes. By 1930 there were still 58 recalcitrant authorities (Cherry, 1974). Progress was undoubtedly slow; in 1933, while more than 9 million acres in England and Wales were covered by town planning schemes, only 94 schemes had been approved, from 50 local authorities. Over vast areas, sometimes in whole counties, town planning was simply not operative. This is not to say of course that some form of building control was not being exercised, through bylaws, but wider, stronger powers, required to combat the countryside disfigurement to which we have referred, were not available.

There were signs of change, however. The Local Government Act, 1929 extended to County Councils the right to share in the preparation and administration of a joint town planning scheme. Furthermore a general interest was shown in regional planning through the operation of Joint Committees. For example, the Manchester and District Joint Committee covered one thousand square miles within four counties, including seven County Boroughs and 89 other authorities. Its plan prepared in 1926 established a general zoning plan for the area. The biggest regional exercise of all was that for Greater London; between 1928 and 1933 Raymond Unwin was Technical Director for the Greater London Committee. His work on open space planning and the establishment of a green girdle for London proved highly significant for future recommendations in this regard. By 1932 no less than 60 Advisory and 48 Executive Joint Regional Committees had been established in England and Wales (Wannop and Cherry, 1994).

Legislative change at the turn of the 1930s was protracted and the final outcome a disappointment to the professionals of the time. In the context of growing concern about rural despoilation, Sir Edward Hilton-Young, a Conservative Private Member introduced his Rural Amenities Bill in 1929, designed to extend planning powers to rural land. It was superseded in 1931 by the Labour Government's Town and Country Planning Bill. The government fell, and the Bill was reintroduced in 1932; it was passed and came into operation on 1 April 1933. There was bitter political controversy over the betterment recovery figure of 75 per cent (Cherry, 1982), some new procedures were considered cumbersome, and there was uncertainty about certain of the clauses, which required interpretation. Above all, the preparation of schemes was no longer compulsory. However, the operation of planning was now 'country' as well as 'town'; the Act enabled schemes to be made with respect to any land, urban or rural, built on or not. In practice, possible claims for compensation against planning refusals militated against much good land planning, though on the other hand the practice of reliance on 'interim development control' while a scheme (often delayed) was in course of preparation did permit the widespread adoption of flexible, short-term regulatory powers, which were often as much as hard pressed local authorities aspired to. By 1938 24 million acres were under some sort of planning control, but only 236,000 acres were covered by the 92 schemes then approved in England and Wales.

While a statutory planning system of a sort took root, in shallow soil, other legislation provided additional related powers. Wasteful, formless ribbon development had long been identified as an objectionable form of building on the rural fringes of cities and the Restriction of Ribbon Development Act, 1935 sought (belatedly) to address the problem. The main provision was that the consent of the highway authority was necessary for access or development within 220 feet of the centre of a classified road. The Act had only limited success, largely due to the divergent interests of highway engineers and planners; the former were content to set ribbons back behind a service road, the latter more concerned to terminate linear building development altogether. There was another unsatisfactory aspect of the legislation in that, if Buchanan's (1958) criticism of it is accepted, the Act 'knocked the bottom out of planning' for years (p. 130). A completely new system of highway planning was introduced, altogether independent of the work that the planning authorities were struggling to do under the Planning Acts.

In the meantime local authorities took what initiatives they could to exercise greater control over rural development. A number of counties, particularly in southern England proved very active by promoting local Bills (Sheail, 1981). The Surrey County Council Bill, given the Royal Assent in July 1931, served as a precedent for other County legislation and indeed for more general legislation, including the Town and Country Planning Act, 1932 and the Restriction of Ribbon Development Act, 1935. Not all local efforts were successful, however; the South Downs Preservation Bill, promoted by East Sussex County Council in 1934, was given a second reading, but fell when referred to a Select Committee for Opposed Bills. On the other hand the Lindsey County Council (Sandhills) Act, 1932 successfully conferred powers to secure the optimum use and management of Lincolnshire sand dunes in the vicinity of Mablethorpe by means of planning controls and land acquisition.

Another successful initiative, by counties around London, concerned the green belt. County councils, supported by other local authorities began to acquire land and properties, before they fell into the hands of speculators, outside the narrow green girdle delineated by Unwin in the Greater London regional planning exercise. In this way a discontinuous ring of green belt estates around London was acquired in a programme of preserving open spaces and farmland by the counties around the metropolitan fringe (Thomas, 1970). In 1935 the London County Council launched its own green belt scheme, offering up to half the cost of acquisition. All authorities stood to gain: London secured a reserve supply of recreational land, readily accessible for its urban population; the counties and their districts also provided recreational facilities, preserved amenities and controlled land supply to prevent urban development. The LCC grant aid scheme made £2 million available. The Green Belt (London and Home Counties) Act, 1938 legitimated the grant offered to authorities outside its jurisdiction. The success of the scheme (from which so much was derived in later planning initiatives) showed that the inadequacy of the statutory planning system could be overcome by political will and a liberal financial approach.

A number of the larger local authorities, counties in particular, made commendable efforts to address the problems of the countryside. Initiatives in regional planning have been mentioned. Councils frequently appointed consultants in the preparation of plans and advisory reports. Throughout the inter-war period professional planning ideology permeated the world of local government. Countryside preservation was a dominant theme in the consultants' reports. Adams, Thompson and Fry in the North East Kent scheme (1930) declared that 'no phase of regional planning is more important than that which has to do with the preservation of amenity and no amenity is more important to preserve than that of the countryside'. The Regional Scheme for Bristol and Bath (1930) by Abercrombie and Brueton advocated a careful zoning plan with belts of land kept permanently open. Adshead (1933) urged control of the Lea Valley in West Essex by zoning. Abercrombie and Kelly (1932) called for the preservation of Lakeland, with further areas purchased and vested in the National Trust and the avoidance of disfigurement by electricity pylons. By the end of the 1930s there was no radical alteration of the statutory system in sight, the Ministry of Health obviously believing that the destruction of amenity through insensitive development could be reasonably checked by existing powers. It is debatable for how much longer that line could have been held, but altogether different circumstances were shortly to force a new approach.

CONCLUDING REMARKS

It is difficult to avoid the conclusion that it was between the wars that twentieth-century rural planning in England and Wales took on its formative, lasting aspects. Lessons from both successes and failures during this period provided the cutting edge which allowed war-time (1939–45) Britain to respond in preparing its plans for the years following the cessation of hostilities. Programmes for the post-war reconstruction of the countryside, the restoration of its economic base, the revivification of its community life, the protection of its scenic qualities, and new provision for outdoor recreation were all predicated on inter-war attempts to deal with these issues.

The nation was of course grappling with enormous social and economic problems, a task made difficult by the absence of any clarity as to which policies to pursue. Prevailing political sentiments in British governments during the inter-war years made it unlikely that land planning or financial support to agriculture would be high on the agenda. The alienation of a land-owning class, either for unrestricted access to private land or for removal of property rights in matters of compensation, was unlikely. In any case, Treasury restrictions on public spending were tight over two decades of sustained financial orthodoxy. But as difficulties mounted and masterly inactivity was maintained on many fronts, the vigour which might have been introduced into public policy initiatives was ceded to other quarters, elsewhere. Professionals became more assertive and more confident in their prescriptive solutions, and special interest groups assumed a role of articulating public opinion, which became very demanding for treasured countryside objectives.

Today's judgement would suggest that opportunities were wasted through overcaution and a timid reluctance to involve government in countryside issues. Let just one example make the point: Dudley Stamp's Land Utilisation Survey of Britain. Set up in 1930, under the auspices of the London School of Economics, this was an independent research organization directed by the geographer L. Dudley Stamp (Stamp, 1946). The objective was to record on detailed maps the use of every acre of land in England, Wales, Scotland and the Isle of Man; a similar survey was later carried out for parts of Northern Ireland. The maps (42,000 of which were destroyed in a *Luftwaffe* air raid in May 1941) presented a fascinating coloured representation of land use in Britain during the early 1930s. The Survey was carried out by unpaid volunteers drawn from universities, colleges and schools. Grants from the Rockefeller Foundation and the Pilgrim Trust came to an end in 1936 and the work was only completed from personal funding. Throughout the venture only negligible material aid was given by official sources, symptomatic of the Government's unwillingness to collect and publish data on the physical environment during this time. The pendulum was to swing with a vengeance as a different view of countryside planning shortly took hold.

5

THE HIATUS OF WAR: PLANNING THE COUNTRYSIDE 1939–1945

In chapter three it was acknowledged that, as far as the countryside was concerned, the Great War was both a period of change *and* a period of continuity. The same can be said of the Second World War. At a first, superficial glance, the plethora of published reports and governmental action which actually followed in the years 1939–45 gives credence to the idea that the period was truly one of radical change. The publication of the trilogy of wartime reports – Barlow (1940), Scott (1942) and Uthwatt (1942) – which directly concerned the future of land planning, and the direct involvement of the State, both local and national, in organizing agricultural production, are perhaps the two most obvious areas of evidence of this idea. The post-war rural world which followed has been characterized by two factors above all – a formal land use planning system and a close and continuous governmental involvement in farming. These are, perhaps, the two great themes of post-war England and Wales – and they seem unequivocally to have their origins in wartime fears and wartime controls.

The idea of wartime as a period of radical social change is, of course, not a new one. In the specific context of the Second World War it is probably most clearly associated with the work of the social theorist Richard Titmuss (1950, 1958). The theme has also been cogently argued more recently by the historian Arthur Marwick (1976), though he would not concur completely with Titmuss's views. There are perhaps three components to the purist view that the Second World War was truly a period of radical change. First, new and original structures and policies were initiated, and in some profusion. Second, these new ways had weak or non-existent pre-war antecedents and they came about *because* of the hiatus of war and were not simply the coincidental grouping of changes which had been long in the making. Third, and this component is perhaps the most important, wartime created a massive change in social relationships with evidence of major strides towards a more egalitarian society.

Titmuss especially argued the latter point, citing for example the views of the media (notably a key leader in *The Times* after Dunkirk) which paved the way for a radical change in social attitudes and thereby led to social reform. This view would now be regarded as misguided – indeed some recent work, for example the volume of essays edited by Smith (1986), would argue the opposite insofar as wartime activities were seen to reinforce old social levels and attitudes. As far as the other components of the argument that there was significant social change are concerned, there is no doubt that here was a remarkable *coincidence* of activity regarding the planning of the countryside during (and immediately after) the years 1939–45. However, as chapter 4 has clearly shown, there was substantial pre-war activity in the development of planning for land use generally, and for both the preservation of countryside amenity and the agricultural economy. The conclusion to which most would be drawn is that the Second World War was certainly a period of hiatus and of major developments in planning the countryside at a time of crisis. Of itself, however, the war was simply the midwife to change rather than the parent.

Before progressing to a consideration of these changes it is important to note the especial way in which rural areas figured in official comment during the Second World War because this undoubtedly reinforced a view of the countryside which coloured later policy. Throughout the war years the countryside formed an important reference point for the British people, and government ensured that its importance was fully emphasized. It was rural England (and not any other parts of the kingdom) which was seen as the quintessential essence of what was being defended – the 'real' England that was under attack. This view figured not only in official propaganda, including speeches by Churchill, but was best seen perhaps in the radio broadcasts of J.B. Priestley. Angus Calder (1991) has referred to this attitude as an appeal to 'Deep England'. It was certainly important in wartime in mobilizing the people to defence, but its influence continued afterwards, for example in views that farmers should be supported since they had rallied round and fed the nation when U-boats threatened, or that rural land and landscapes should not fall unthinkingly to the depredation of urban growth when they had been successfully defended against Hitler.

These two points – the true significance of the War and the image of the rural emphasized in wartime – come together. The images of 'deep England', of the 'real' Englishman being a *rural* Englishman and of the threat to rural landscapes, are all old images and long pre-date the outbreak of war. Indeed some of them figured strongly in the 1920s and 1930s when economic pressures presented different threats and crises. But the war was genuinely a hiatus and provided a stark and immediate focal

Farming on the brink of war. A threshing scene in the Thames Valley in September 1939. Note the large number of men working who have yet to be called up, and the children watching ready with their newly-issued gas masks. (By courtesy of the Rural History Centre, University of Reading)

point for their significance. The thoughts were not newly-created, but they were given new urgency and emphasis by the war and were apposite to the time.

A further point must be borne in mind in considering wartime development: the country had been there before a generation earlier. As will be seen, several actions which are conventionally seen as 'wartime development' had their origins in the 1930s as war seemed to get nearer and arrangements against the eventuality were put in place. The Second World War contrasted greatly with the First when, of course, there had been no precedent for a continental war and the impact upon the rural areas was scarcely thought of. British agriculture in 1914 was, relative to the depression of the late nineteenth century, moderately prosperous (Howkins, 1991). The position in 1939 after twenty years of farming recession was very different. The land was in a poor state and food supplies from home sources were at an all-time low. This situation of course lent extra emphasis to the view of the critical importance of the countryside and the farming industry.

Three main themes are recognized in this chapter. First it is necessary to plot the immediate effect of planning for the wartime situation as policies were adopted to defend and feed the nation. Second there is a need to provide a picture of the rural community in wartime as it was influenced by rapid and major social and economic pressure. Third it is necessary to analyse the great effort made towards rural reconstruction in the post-war world. Even though the 'war as cause for social change' argument may now be rejected, there can be no doubt as to the scale of wartime activity in rebuilding the economy and society of the countryside, nor of its significance for many years after.

PLANNING FOR WARTIME

The response of the farming community to the need to feed the nation in the face of German blockade is part of the rural folklore of the twentieth century. By any standards the response was a remarkable one and for it, some might argue, the farming community exacted a reward from a grateful post-war generation which was to be paid back over more than forty years.

By the late 1930s more than 20 million tons of food were being imported annually while home agriculture could barely account for 30 per cent of self-sufficiency. In England and Wales the total area of land in agricultural production in 1939 (24.6 million acres) was 2.3 million acres fewer than had been the case at the end of the First World War. The area under crops had fallen by more than 30 per cent, with reduced areas under wheat, barley, oats, potatoes and root crops. Livestock numbers, particularly dairy cattle, had increased somewhat over the same period but this was mainly a response to formerly cropped land falling to grassland as farmers cut costs and reverted to livestock farming.

By 1945 the picture was very different. While the crude area of land in all agricultural production had in fact declined a little from the 1939 figure of 24.6 million acres to a 1945 total of 24.3 million acres, mainly due to the transfer of land to use by the armed services for airfields and the like, the arable acreage had increased substantially from 8.9 million acres to 14.5 million acres with particularly large changes in the area under cereals and potatoes (Hammond, 1951; MAFF, 1968). While much of this increase was made possible by the ploughing up of permanent grassland, this was not accomplished, contrary to some warnings, at the expense of milk supply. While numbers of other livestock fell, as grazing land was ploughed, those for cattle increased, fed increasingly on ley grasses and fodder crops which replaced imports. Milk, seen as a crucially important item in the national diet, especially as far as mothers and children were concerned, was thus maintained in supply and even increased slightly by the end of the war.

Overall, whichever measure of agricultural output is chosen, the record of British agriculture in meeting the challenge of increased home supply was a remarkable one.

How was this accomplished? At the outset it is important to realize that the agricultural effort during the Second World War, in strong contrast to that in the First, benefited in several ways from preparations and actions made before hostilities commenced. As already noted, the 1920s and 1930s had seen the gradual involvement of the State in agricultural production through price support and subsidy, marketing boards and the like. Moreover, in some cases at least, agricultural science had begun to provide an important foundation for a modern agriculture which could rise to the challenge of increased production. Most obviously, perhaps, this could be seen in the work of Stapledon in grassland improvement (Stapledon, 1935; Waller, 1962). Mention of Stapledon leads on to a further point regarding pre-war preparation, for he had instituted from 1938 a survey of the grassland resources of Britain which was to be a key element, along with Dudley Stamp's Land Utilisation Survey, in the ploughing-up of poor quality pastures and their eventual improvement.

The 'Plough-up Campaign', as it became popularly known, was perhaps the most visible effect of the war upon the rural landscape. Land which had been under grass for generations was turned to crops, whether on the heavy clays of the English Midlands or the light brashy soils of the Cotswolds and the Downs. It should be appreciated that the process was, in fact, not one single campaign as such, for each new growing season provided the opportunity of a new target for farmers to plough new land. Thus, as the classic account of wartime agriculture by the doyenne of British agricultural economists Edith Whetham (1952) recounts, the first year of war saw the setting of a target of 2 million acres to be ploughed, while subsequent years saw additional acreages of 2¼ million acres (1940/41), 1¼ million acres (1941/42) and 1 million acres (1942/43).

Activity on this scale presumed strong local control and this was essentially provided by the War Agricultural Executive Committees (the 'War Ags'). These were county-based committees of local farmers and landowners whose job was to supervise the national agricultural policy, including the plough-up campaign, and to maintain and improve the quality of farming in their area. As chapter 3 noted, such committees had been created during the First World War, but only in 1917 (Dewey, 1989). The 'War Ags' were another example of how the Second World War saw rather better preparation. They had in fact been quietly re-created as early as 1936 and a structure of membership put in place in the counties (even though many members did not at that stage know of their appointment!). Their existence was publicly announced in May 1939 and they were thus ready with contingency plans for the outbreak of war four months later.

One of their first acts was a rapid survey of all farms in their area as a basis for gauging the potential for increased production, which survey was followed up in much greater detail in the period 1941–43 as a National Farm Survey.

The War Agricultural Executive Committees were a crucial element in the prosecution of the farming war effort, and indeed in modified form they continued in an advisory capacity for many years after the war. They clearly had what in retrospect appears as virtually dictatorial powers especially insofar as they could order farmers to improve the quality of their farming, require land to be ploughed or fertilizers to be employed. In extreme circumstances they could take land away from an inefficient holder and give it to others to farm better. From September 1939 to March 1945 some 443,000 acres of land in England and Wales were requisitioned and the tenancies on nearly a quarter of a million further acres were terminated (Whetham, 1952).

The overall impact of the war upon the farmed landscape was very substantial. It amounted to nothing less than a massive turn around in agricultural practice and in the appearance of farms and the rural areas generally (e.g. Stamp, 1947, 1962). Equally it saw a revolution in national policy towards agriculture. While in no way gainsaying the developments

The wartime 'plough up' campaign. A scene near Hampton Court, Surrey in the early years of the Second World War. (By courtesy of the Rural History Centre, University of Reading)

regarding State involvement in the 1920s and 1930s, the wartime years consolidated and made comprehensive the foundations of post-war farming policy. As chapter 6 will show, the cessation of hostilities was perhaps less of a hiatus in policy terms than might have been expected; Whetham's 'brief history of British farming during the decade of the Second World War' quite properly extends to 1949 and therefore encompasses the legislation and policy formulation of the post-war Labour administration.

THE RURAL COMMUNITY IN WARTIME

No Census of Population was taken in 1941, the first break in the decennial pattern since the first count of 1801. Under normal circumstances it is certain that, had a Census been carried out, it would have shown the continuing loss of people from the countryside to the towns which had been happening since the middle of the nineteenth century (Saville, 1957). But wartime was not normal, and a Census would have shown the consequences of a substantial redistribution of the population, mainly from urban to rural areas in a curious foretaste of the counterurbanization movements discernible twenty-five or thirty years later. Together with the movement of service and other personnel around the country, these movements of official evacuation and personal migration made the war period one of major demographic upheaval.

As was noted at the outset to this chapter, critics are now very much more circumspect as to how far these *demographic* movements were significant in engendering radical changes in *social* attitudes and values. Nonetheless in the short term the demographic and consequent social upheaval was of major proportions and affected the lives of most of the population in some way.

Foremost among these changes which affected the countryside was evacuation, mainly of children but also of some women and others in what were judged as 'priority classes'. Although plans had been made for the evacuation of vulnerable urban areas during the 1930s, when evacuation eventually started on 1 September 1939 it was carried out amidst some chaos and disorganization. Children and women were scattered widely to rural areas covered by no less than 476 billeting authorities, families were split up and children traumatized by long journeys to strange and secret destinations (Macnicol, 1986). In all fairness, the scale of the evacuation was of enormous proportions. Titmuss's (1950) official history of social policy suggests that over just a few days in early September 826,959 unaccompanied schoolchildren, 523,670 mothers with pre-school children, 12,705 expectant mothers and 173,000 others were transported from the cities.

For a short while at least the effects upon many small rural communities and individual households were dramatic. Official reports noted the clash as urban met rural often for the first time. A major focus of interest seems to have been the health and behavioural condition of the urban children. By contrast with their new rural contacts they were frequently reported as ill-nourished, dirty and verminous, with many adverse comments being made on their behaviour and manners (Macnicol, 1986). Late revaluations of these judgements would moderate their tone and point out an obvious class bias since they were invariably the comments of middle class rural leaders involved in the evacuation process who previously had but little to do with the children of the urban proletariat.

This first evacuation was, however, remarkably short-lived. It appears now akin to a panic measure as war approached. As German bombing did not occur as expected, the evacuees soon began to drift back to the towns. By the beginning of 1940 it was reckoned that only 20 per cent of the original evacuees remained in the countryside. There were, however, to be later waves of evacuation, notably during the Blitz (1940–41) and again in 1944 when the V1 and V2 rocket attacks began. Moreover it is estimated that upwards of two million additional people evacuated themselves and were not part of official movements. Official and unofficial movements from town to country probably, if Titmuss's estimates are taken, affected no fewer than six million people throughout the war.

The significance of evacuation has generally been seen through the eyes of the urban population. Here there seems general agreement that it was a major and transforming experience for a whole generation of urban children (Macnicol, 1986; Ward, 1988), though commentators would probably now reject the Titmuss thesis that the evacuation and its lessons were a direct cause of a post-war social reformist movement which led inevitably to the creation of the welfare state. The impact upon rural society is far less well recorded but for many country people it must have been a first introduction to lifestyles in the big city. Certainly over the whole period of the war the social disruption caused by evacuees and other temporary visitors, coupled with the call-up of the menfolk, must have broadened the outlook of many rural communities.

This is not to say that social divisions within rural communities were necessarily broken down. Indeed it seems likely that, if anything, the old social order was emphasized by the process of evacuation. Leaders within the local community had their status reinforced as billeting officers, air raid wardens, special constables and the like. The new organization was automatically put in the hands of those who already had status and position.

A second major change in rural communities brought about by the wartime situation concerned the agricultural work force. There was on the one hand something of a conflict between the needs for farm labour to deliver more food for the nation, and on the other the demands of the armed services, munitions factories and the like. The compromise was achieved by the exemption of many farm workers under the terms of the Military Training Act of 1939 whereby those aged over 25 (later reduced to 21) were exempted from call-up. Nonetheless by March 1940 around 50,000 workers had been lost to agriculture (Armstrong, 1988), going either to the armed services or to work in other industries related to the war effort. Towards the end of the war, the figure of 'skilled men' lost from agriculture was put at 98,000 by a government White Paper issued in November 1944 (Ministry of Information, 1945).

Despite this loss, the bulk of farm work done during the war years remained with the traditional labour force, augmented as noted below. In fact for the farm worker wartime brought with it some real progress in conditions, notably with regard to pay. Prior to the war, farm wages had generally been set locally and invariably at levels well below that of comparable occupations. From 1940 a national minimum wage was set, linked, at the farmers' insistence, with the price for agricultural products – a foretaste of the national involvement in both farm wages and prices which was to be a notable feature of agricultural policy for the next forty years. So, just as farm production was supported and the prices received by farmers rose, so too did workers' wages. The average minimum weekly wage in 1939–39 was 34 shillings and 8 pence (*c.* £1.70); by 1945–46 this had more than doubled to 72 shillings and 2 pence (*c.* £3.60), though the cost of living over the same period had only increased by half (Ministry of Agriculture, Fisheries and Food, 1968).

Over the whole period of the war, the agricultural workforce grew after years of gradual decline. From a low point at the outset of war, of around 580,000 full and part-time workers, by 1945 the total had risen to 766,000. Some of this expanded number came from the re-employment of older workers, but the bulk of the increase came from two other sources – the Women's Land Army ('land girls') and prisoners of war.

The development of the Women's Land Army, which had been created in 1939, just before the outbreak of war, was very much in the hands of the redoubtable Lady Denman, the central figure of the Women's Institute movement. Early recruitment was slow but grew to 87,000 by the autumn of 1943 (Armstrong, 1988; Sackville-West, 1944), virtually all recruited from the towns and from non-rural backgrounds. The social impact upon rural communities, and especially upon an industry that had for long been male-dominated, must have been substantial, though inevitably the evidence is anecdotal.

The employment of prisoners-of-war on farms tended to come later in the war, especially after the success of the North Africa campaign and the

Training the Women's Land Army at the South Eastern Agricultural College, Kent. Trainees, who came from all walks of life, were given a month's training; here they are being instructed in the use of the horse-drawn cultivator. (By courtesy of Wye College)

capture of many Italians. Despite some initial uncertainty about allowing them the 'freedom' to work on farms, there were by mid 1944 more than 50,000 prisoners of war engaged in agricultural work (Armstrong, 1988). Some of those, both German and Italian, remained behind after the war, often marrying local women and becoming integrated into the local community.

Beyond the influx of evacuees and the appearance of new workers on the land, the war obviously affected the day-to-day lives of rural people. Key workers, such as tradesmen and schoolteachers, were called away to the services and the normal pattern of life inevitably disrupted. Social relationships were affected as families were split up and new people came to live in the village. Contemporary accounts inevitably emphasize the change, disruption and uncertainty which wartime brought. But beyond these obvious effects two related themes stood out. Firstly the concern of all involved was that the life of the countryside should continue as much as normal. Of course the times were abnormal in every particular, yet there was clearly a strong desire to pursue a feeling of normality, that the countryside was the same as it ever was.

Secondly it is very clear that, whatever might happen after the war, the old-established social and class relations which had structured the life of the countryside were very firmly in place. Indeed the rural war effort itself was a clear manifestation of the class structure. The choice of 'key' people to serve on the War Agricultural Committees or to make evacuation and billeting arrangements was an easy one. Wartime social relations were no different in that respect from peacetime. Farmers and tradesmen, middle class bourgeois and employers, were seen, or saw themselves, as the natural leaders. Organizations such as the Women's Voluntary Service, the Women's Institutes and the Young Farmers's Clubs certainly adapted their activities to meet the national emergency, but they operated just as they had always done. Accounts of 'the countryside at war' (e.g. Grant and Maddren, 1975) emphasize how the rural class structure was actively maintained. Perhaps one of the best personal accounts of this type is the 'letter' sent to American friends at the outset of war by the novelist Margery Allingham and published in 1941 as *The Oaken Heart*. Her account of the onset of war in her home village of 'Auburn' (in reality Tolleshunt D'Arcy in Essex) highlights many of the factors noted above: the invasion of urban evacuees, their curious manners and doubtful hygiene, the resilience, humour and common sense of country people in the face of the earth-shattering developments on the international stage. Above all there was the implicit leading role taken both by the author and her family and her middle class neighbours in organizing the village to deal with wartime conditions. Wartime did not change these established relationships; it just accentuated their significance.

PLANNING FOR POST-WAR RECONSTRUCTION

At the same time as major developments were taking place within agriculture and the wider economy to ensure the direct contribution of the home front to the war effort, a second agenda was also being enthusiastically followed: reconstruction after the war. In every area of activity, not just the countryside, there was a positive welter of books, reports and the work of committees and conferences, which appeared particularly after the immediate threat of invasion was past and which focused upon how the country would be planned after hostilities. A strong feeling that in the First World War mistakes had been made and that afterwards opportunities had been missed because of a lack of planning, made many people determined that it would not happen again. The canvas was broad – not just agriculture and land use, but the whole system of education, social welfare, the health system, industrial development and economic growth were seen as grist to the planning mill. As Cherry (1974) has said 'The idea of planning swept through virtually every aspect of Government activity; there was the determination to make plans for post-war years, to set our national house in order, to grapple with the problems we had shelved.'

The countryside and how it should be planned seems to have been a particular focus for exhortation which emphasised the need above all to plan. In agriculture eminent scientists and leaders such as Sir Daniel Hall, Dr C.S. Orwin and Sir George Stapledon were joined by many others in emphasizing national planning, controlled marketing, and the application of science. Hall's *Reconstruction and the Land: An Approach to Farming in the National Interest* (1941) and Orwin's *Speed the Plough* (1942) provide classic examples. The broader countryside was the focus of several books – Thomas's *The Changing Village: An Essay on Rural Reconstruction* (1939), Orwin's *Problems of the Countryside* (1945) and McAllister's collection of essays *Homes, Towns and Countryside: A Practical Plan for Britain* (1945) are all very much in the mould. Hardy (1991) has noted the contribution of the Town and Country Planning Association during this time as a forceful pressure group for the increased role of planning. In consecutive years, the Association's annual conference concentrated upon the future of the countryside. In 1942 industry and rural life were discussed (Newbold, 1942), while country towns were the focus for the 1943 conference (Baron, 1944).

At a superficial glance this undoubted enthusiasm for planning seems to identify the war years as a period of very radical change in the management of the economy as a whole and of the countryside in particular. The profusion of books and reports referred to above was matched by an equal measure of official activity: Acts of Parliament, White Papers, Departmental Reports etc. Yet the coincidence within a few years of so much activity was arguably no more than that: the coincidence of feverish activity brought about by wartime emergency which brought to a head ideas and actions which had been in the making long before. Certainly this argument is borne out when the two main areas of activity which focus on the countryside – farming and the control of land use – are considered.

As chapter 4 showed, the inter-war years saw significant moves towards State influence and control in farming, notably with price supports for various commodities and the creation of marketing boards. In land use planning and rural conservation too, wartime activity had been presaged by important developments in the interwar period. A clear instance can be seen in the case of the 1940 Report of the Barlow Commission which had been set up in July 1937 'to inquire into the causes which have influenced the present geographical distribution of the industrial population of Great Britain . . .' The recommendation of the Commission that there should be a central planning authority (rather than individual boroughs deciding to take planning powers for themselves) began to be realized in the passing in 1943 of the Minister of Town and Country Planning Act, of further planning acts in the same year and the Town and Country Planning Act 1944. It is apparent that the effective beginnings of comprehensive urban planning date from this time.

As far as a more narrowly defined countryside planning was concerned, the central feature of the wartime years was undoubtedly the publication in August 1942 of the Report of the Committee on Land Utilisation in Rural Areas under the chairmanship of Lord Justice Scott. The Scott Report was, much more than the Barlow Report, essentially a child of the war. The Committee had been set up by Sir John Reith in October 1941 during his relatively short but dynamic period as Minister of Works. Reith's original intention seems to have been for an investigation into rural industries and in particular into which industries were especially suitable for location in the countryside. In the end the Committee was given a broader brief which they stretched to the full, a brief reflected in its now well-known, if ill-written, terms of reference:

> To consider the conditions which should govern building and other constructional development in country areas consistently with the maintenance of agriculture, and in particular the factors affecting the location of industry, having regard to economic operation, part-time and seasonal employment, the well-being of rural communities and the preservation of rural amenities.

Scott was an old hand as far as planning matters were concerned and he was joined as vice-chairman by the geographer Professor L. Dudley Stamp, director of the Land Utilisation Survey of Britain. The planner Thomas Sharp acted as one of the joint secretaries together with Basil Engholm who was later to go on to be head of a new Land Use Division in the Ministry of Agriculture. Wibberley (1985) was later to argue that much of the tone of the Report was set by Stamp and the two secretaries. Indeed, the role of Stamp in the writing of the Report is very clear to anyone with a knowledge of his other activities and writings. There is a strong emphasis on the key role of land use survey as a precursor to planning, and the Report, in common with much geographical writing in the past, is long on description and short on rigorous analysis. Some premises, notably the 'natural' dominance of agriculture as a land use in the countryside and the belief that farming was an inevitable and benign mainstay of landscape protection and nature conservation, were accepted uncritically. Anyone reading the views of the majority of the Scott Committee a half-century later would be amazed at some of the assumptions, especially regarding the future of agriculture. They were convinced that what was required was 'the continuance and revival of the traditional mixed character of British farming' (para. 46). More dramatically, they felt able to state categorically:

> In our opinion a radical alteration of the types of farming is not probable and no striking change in the pattern of the open countryside is to be expected. (para 46)

Sharp's influence upon the Report, too, is clear. As Stansfield (1981) has shown, this 'somewhat difficult man' (Cherry, 1981) had very clear ideas about the need to separate town from country. The essential message in the Scott Report was exactly in tune with this belief. From the viewpoint of half a century later, the Report appears both dated and misguided. At the time it did not appear so and the Report was generally well received. As Wibberley (1985) has said, it seemed as though Stamp, Sharp and Engholm had asked themselves the question 'What would the ordinary person, concerned about the countryside, like to hear at this time?' – and had then set out to answer it.

Its viewpoint and recommendations would be enshrined in post-war attitudes, legislation, policy and practice over a field much broader than that just related to the planning of land. For example, the Committee felt strongly that the quality of rural life should be brought up to the levels of the urban population and at no extra cost to the individual rural dweller. Thus:

> The supply of electricity is an essential service which in due course should be available in the house of practically every citizen in town and country alike, at no higher price to the consumer in the country than in the town. (para 165)

In this could be seen the forerunner of the uniform pricing policies for the major services of power, water and telecommunication (and indeed of a whole local government financing system) which was to be built into the policies of the new nationalized industries.

The Scott Report occupies a pivotal position in rural planning history because it elaborated a major theme which would be present for many years to come: a land use planning system should have as its prime duty the objective of protecting agricultural land. That is, planning should be seen as essentially operating within an urban context to control development and keep the countryside free for farmers to go about their business. This theme had been heralded in the Barlow Report but Scott went further insofar as it provided for the first time a formal and authoritative statement of the link between land use planning and the farming industry. Previously, though state planning had become gradually more important in both spheres since very early in the century, the two areas had been considered separately. The Scott Report linked the two primarily by arguing that planning was in the main about protecting farmland and that, in particular, farming should invariably be considered to have a prior claim to land use unless competing uses could prove otherwise.

This notion of a prior claim, enshrined in the 'onus of proof' argument, is worth repeating in detail as stated by the Scott Committee (para 233).

> *The Onus of Proof* – Land which is included in one of the categories of good land should not be alienated from its present use unless it can be clearly shown that it is on balance in the national interest that the change should be made. The same applies when the question relates to land which, though of indifferent quality, may be an essential part of a well-balanced farming unit.

> We attach real importance to the *onus of proof*, whether a decision is being reached in a matter of wide national importance, or in a purely local case. It is not merely that agriculture is in possession, actually or potentially, and that possession of itself puts the burden on the applicant who seeks to make the change, and, in the case of building, to bring to an end for ever the present form of utilisation. It is rather the method of approach to the problem. Where the land is of a good agricultural quality and there is no dominant reason why there should be constructional development, the task of the Authority is simple – its answer will be 'No!' But in the case of some of the intermediate qualities of land, especially where *pros* and *cons* are at all evenly balanced, or other sites are offered to the applicant as alternatives (and it is not easy to gauge their suitability in all such cases) it would be of very general assistance to all persons likely to want sites for construction, as well as to the owner of the agricultural land, and the Minister of Agriculture himself, if it were common knowledge that agricultural sites would not be handed over unless a clear case of a national advantage was made out.

Fifty years on, in a world in which agricultural land has been argued as being in surplus, and where the agricultural industry no longer holds the centrality of esteem which it once did, this argument seems specious. And yet it was substantially supported at the time, and in general terms held sway with regard to planning decisions at least until the 1980s. It provided the *leitmotif* for the Town and Country Planning Act 1947 and laid the foundation for the influential role of the Ministry of Agriculture and its local officials in the determination of many local planning disputes in the post-war years. A small indication of the Report's long-standing influence in the post-war planning of rural areas can be gauged from the fact that it was reprinted as late as 1967.

The onus of proof argument was, of course, only one specific aspect of a general view which was powerfully held during the War. Many people, especially those of influence and with power, argued that the response of farmers to the war effort should be rewarded once hostilities ceased. The signatories of the majority report of the Scott Committee certainly saw it that way and they undoubtedly helped substantially to create the climate in which the post-war farming industry would be nurtured.

This is not to say that this majority view was universally held. An editorial in *The Economist* at the end of August 1942 put forward a warning which in retrospect seems prescient:

> Afterthoughts on the majority report of the Scott Committee bring no added ray of sunlight. Its plea for the preservation of the countryside by traditional farming is, above all, well meaning. But it mirrors an attitude which regards farming as Service and industry or trade as Sordid Business. The industrialist, in this light, makes wartime profits; but the farmer, a horse of quite a different colour, Does his Duty. This sort of sentimentality can be very dangerous if it is taken too seriously. In spite of these romantic and quite ill-founded distinctions, a main aspect of wartime farming is the material gain

> which accrues, quite rightly, to the farming community. The present abnormal farming conditions, due primarily not to any motive of service, but to the attempted German blockade, cannot be isolated from the general structure of prices and values. There is already pressure upon the government to maintain the present artificial farm incomes and agricultural land values when the blockade is over – apparently by keeping the blockade on by peacetime means. The argument is that, because Duty has been done, farmers have a right to maintain wartime profits in peacetime and landowners to keep the wartime betterment in the value of their land after the war is over. To accept this argument would, in the first place, be perilous nonsense. It would not only reduce Britain's standard of living; it would also aim deliberately at eliminating Britain as a first class economic power, which Britain only is by virtue of its ability to make purchases from abroad on favourable terms. In the second place, it would be vastly unfair. Farmers have no more right than munition makers to keep in peacetime the windfall gains of war. They, like every other producer, must stand or fall after the war by the competitive efficiency with which they can employ national resources. The test must be, not the inflated money values of wartime, but the ability, over a period, to produce and sell at competitive prices. Farmers will indeed have a right to some protection against disastrous *fluctuations* in prices; but no right at all to the protection of wartime standards and wartime methods at the expense of the entire community. A basic aim must be a living wage and decent living conditions for rural workers; and this can only be achieved in a farming of a size and of a kind capable of paying its way in face of competition from other pursuits at home and agriculture overseas. It is too often forgotten that the land of this country is individually owned, and that the demand for the artificial maintenance of wartime values or for compensation for their loss is, in fact, a demand for a gigantic subvention to the owners of landed property – an extraordinary piece of special pleading under the excuse of Duty Done.

The Economist, of course, saw through much of the wartime talk about the living standards of the rural work force for what it really was – a covert demand to maintain the profits of landowners. This should not be surprising; after all a good proportion of politicians and society leaders, both national and local, were themselves landowners and farmers. Where expert and influential opinion which was rather less influenced by personal gain was forthcoming, it is interesting to see that the expected returns from land rent which would come from a post-war supported agriculture were seen as property belonging to the State and not to private landlords. Thus Hall, in his treatise of 1941 already quoted, argued 'the first requisite of a constructive policy is the acquisition of agricultural land by the State.'

These suggestions of land nationalization (coming generally, it should be noted, not from hard-left politicians but from liberal-minded academics and scientists) were of course only the agricultural manifestation of the broader question of land betterment which was addressed by the Uthwatt Committee. Farming fortunes apart, the need for the State rather than the individual to benefit from enhanced revenues falling to those who happened to be the 'chance' owners of land was much debated. As C.S. Orwin said in his paper to the TCPA Conference in 1942.

> The issue is a simple one. Does the chance possession of land, or its deliberate acquisition in the hope of capital profits, entitle the owner to exploit the needs of the community for fresh air, for housing, for transport, for recreation, or for any other public purpose, for his own financial advantage? . . . Acquisition of the freehold of the land by the State at valuations based upon its present use, not upon its prospective value, must be accepted as a prerequisite of planning control. (Newbold, 1942)

Rejection of the view of the primacy of agriculture, and in particular of the onus of proof argument, was in fact voiced from within the ranks of the Scott Committee itself. In a Minority Report which in its own way has become as well known as the Majority Report, Professor S.R. Dennison, a young economist from University College, Swansea, seconded to the War Cabinet Secretariat, effectively rejected the key premises and conclusions of his colleagues. He argued strongly in a cogently-written Report that, far from insulating farming and its labour force from urban pressures, the countryside should be encouraged to engage with the mainstream of national urban life. Moreover, he felt that only in this way would the countryside see the level of economic growth which would provide the revenue to improve rural living standards and maintain public services. The Scott Committee had argued that 'the maintenance of agriculture [and the improvement of rural services] . . . will in themselves have the effect of reviving country life and bring about an improvement in the physical and social standards of country areas' (para 199). This was emphatically rejected by Dennison, who saw improvement only coming from rural land and rural resources being put to their most economically rewarding use. It closely followed that he had a much more positive view of rural industrialization which would provide, not unwelcome and destructive competition for farming, but opportunities for betterment for an ill-paid rural population.

These views would today be seen by many people as no more than a sensible and acceptable way to approach questions of land competition and resource use. The farming fundamentalists who remain, and whose attitudes were effectively supported by the Majority Report of the Scott Committee, would certainly continue to reject Dennison's approach. In passing it might be argued that some nature conservationists have unwittingly taken on the onus of proof argument and would argue as uncritically for conservation objectives having 'possession' above competing development pressures. But at the time, Dennison's view was generally ignored and the arguments of the Majority Report, and in particular its attitude to development on farmland, set the tenor of rural planning for the next thirty or more years (Curry, 1993).

The Scott Report, under its wide and rather vague terms of reference, also considered questions of landscape amenity and access to the countryside. The enthusiasm for hiking which had developed during the inter-war years (chapter 4; Cherry, 1975; Blunden and Curry, 1990) was reflected in the Committee's suggestion for a Footpaths Commission to encourage local planning authorities to give rights of way some clearer statutory provenance. More importantly it emphasized the need for the creation of National Parks and recommended that a central body be created to oversee their development. The creation of a system of National Parks and the development of an organization to oversee them was to be a major theme in the story of rural planning in the early post-war years. As such, and for the sake of continuity, it is appropriate that the wartime progress towards this end is told in detail elsewhere. Chapter 8 takes up this story and places it in its wider context of growing demands for countryside recreation and official attempts to plan for this development.

CONCLUSION

This chapter has shown that in all the main areas of activity which concern us, farming, the control of land use and development, and the concern about access and National Parks, the events of the period 1939–45 had virtually all been substantially presaged by earlier developments. This is in no way to deny the very substantial amount of activity in just five short years in all these areas. The point is that wartime gave two ingredients to rural planning – it provided a *focus* and added an *impetus*. The focus centred around a heightened feeling that the countryside was truly important. The nature of the importance differed between different social groups and interested parties. For some the importance was essentially practical insofar as, for example, the countryside was seen as the key resource for food supply. For others, particularly those concerned about access and urban recreation, the importance was more ideological if not directly political. But all of these came together especially when fostered by a propaganda both stated and implicit which emphasized the value of 'deep' countryside.

The impetus came essentially from two sources. The most obvious was the emergency of the times and the threat of invasion and defeat. The second may perhaps be found in the activities of powerful, active and thoughtful men and women. It is difficult to understand the countryside of the Second World War without reference to such individuals as John Reith, Lord Justice Scott, Stanley Dennison, Dudley Stamp and John Dower.

6

PLANNING FOR AGRICULTURE AND FORESTRY IN THE POST-WAR WORLD

In seeking to understand the role of agriculture and forestry in the planning of the countryside in the second half of the twentieth century, there is a temptation to hark back to the situation at the end of the Great War to find useful comparisons. On the surface the parallel appears clear: a rural economy which had suffered a generation of economic depression to be followed by the crisis, and opportunity, of a continental war where increased home food production became paramount. And to match this, there was a set of government promises regarding full support for farmers and landowners, enacted in the desperation of wartime to boost production, but apparently to be continued after victory.

The superficial similarity of situation belies the reality. Put simply, as chapter 4 made clear, the promise to the farming community was broken in 1921. In contrast, though farmers often appeared grudging in their recognition, after 1945 the promise was effectively kept. Wartime encouragements to increased production were carried forward into post-war legislation and policy which guaranteed prices at levels agreed annually in direct negotiations between government and the leaders of the farming industry. A system of grants was progressively created after the war, and particularly so after 1957, to encourage farmers to invest capital in new buildings and equipment and to modernize their farming practices. An equivalent system of support was created to encourage afforestation to increase the home production of timber. By way of further support, a very substantial extension service was created which provided free advice to farmers and landowners and which, through State research stations, the universities and funded research programmes, provided the scientific foundation for the new agriculture (McCann, 1976). And to complement the supportive agricultural policy, there was of course the parallel

Post-war agricultural development. Science meets practice in a demonstration of a new mobile sheep footbath at a meeting of agricultural scientists in Kent. Note the ubiquitous 'Fergie' tractor, developed by Harry Ferguson and, through its use of hydraulic systems, a key element in the post-war mechanization of farming. (By courtesy of Wye College)

development of a land use planning system which had a major objective to control urban growth and keep the countryside as free as possible of incursions which would detract the farmer from his task of producing as much food as possible.

The post-war consensus regarding the importance of agricultural production, and all that flowed from this understanding, from massive State support to the high esteem within which farmers were generally held, lasted for a long time. There were naturally times, as in the mid-1950s, when government attempted to rein back the system when the high cost was realized and throughout the early years after 1945 the public view of farmers included a wry recognition of 'feather bedding'. But by and large the farming industry, and to a significant extent the forestry sector also, enjoyed after 1945, a period of generally settled support which lasted until at least the early 1970s. Indeed, as will be seen, policy statements even as late as 1979 (MAFF, 1979) expounded a view of the critical importance of increased agricultural production which would not have sounded out of place thirty years earlier.

STATE SUPPORT FOR AGRICULTURE

The Labour Party which came to power in 1945 had set out a radical scheme for agriculture in a wartime position paper ('Our Land') published in 1943. When legislation finally came in a suite of Acts in the period 1946–1948, the policy of the party in power was seen to be rather less radical than the 1943 statement had intimated. Questions of land nationalization had noticeably been avoided (Whetham, 1952). But in its own way the policy approach set out by the legislation of the late 1940s was radical enough.

As a prelude to the centre-piece of the 1947 Agriculture Act, a Hill Farming Act was passed in 1946 which introduced for upland areas a system of grants for land improvement, buildings and infrastructural improvements such as electrification and roads and which continued the system of headage payments for hill sheep and cattle which had been introduced during the War. In the same year, a comprehensive country-wide extension service, the National Agricultural Advisory Service (NAAS), later to become the Agricultural Development and Advisory Service (ADAS), was also put in place as a major element in the drive for farming efficiency. This was linked at the county level with Agricultural Executive Committees. These committees, made up of local farming leaders and staffed by the new advisory service, were in effect a direct continuation of the War Agricultural Committees. They retained many of the draconian powers of their wartime counterparts, including supervision and even dispossession, though that was rarely used. The validation of their activities very much relied on their link with the advisory service which was soon recognized as proficient and beneficial. As such their actions were generally welcome or at least tolerated by the farming community.

The major Agriculture Act of 1947 confirmed the government's commitment to agricultural support (Winegarten and Acland-Hood, 1978). While ostensibly clear, its main objective in retrospect seems vague and uncertain, stating the intention of:

> promoting and maintaining, by the provision of guaranteed markets and assured prices . . . a stable and efficient industry capable of producing such part of the nation's food and other agricultural produce as in the national interest it is desirable to produce in the United Kingdom, and of producing it at minimum prices consistent with proper remuneration and living conditions for farmers and workers in agriculture and an adequate return on capital invested in the industry.

The Act created a mechanism for guaranteeing prices bolstered by deficiency payments for the major agricultural products which were to be agreed in annual price reviews negotiated with the National Farmers' Union. The promise of the Act to produce food at 'minimum prices' was more apparent than real for although the price of food in the shops was ostensibly kept low, the real cost of such support was inevitably borne by the consumer in the guise of taxpayer.

A final Act, the Agricultural Holdings Act of 1948, completed the post-war legislative package by in effect allowing a tenant farmer to have a life-long tenancy and also by making provision for compensation in the event of the cessation of the tenancy. In this way it was hoped that tenant farmers, who at that time farmed around two-thirds of the agricultural area of England and Wales, would be given the confidence and continuity of occupation which would encourage them to invest in fixed capital and in new farming practices. This security of tenure remained throughout the whole post-war period, and was indeed strengthened in 1976 to extend to the farmers' successors. While justified, as noted above, on the grounds of economic investment, there were early doubts as to its wisdom. As Whetham (1952) stated:

> Doubts have already been expressed whether this absolute security of tenure to all except the conspicuously inefficient is really in the best interest of the farming community. No landlord can now remove a tenant of only moderate standards in order to give a chance to the young man of promise and energy.

Subsequent legislation in the 1950s and 1960s combined this broad theme of State support for agriculture with a further Agriculture Act in 1957 which tried to stabilize guaranteed price levels and which also introduced a new scheme (the Farm Improvement Scheme) which encouraged land improvement in the uplands. The Scheme also made grant aid available for the construction of buildings on all farms – a provision enthusiastically taken up by farmers in the lowlands. The Agriculture Act 1967 provided grants to encourage the amalgamation of small agricultural holdings and created a new system of grants for land improvement, especially on hill land where productivity increases were seen as especially necessary.

Throughout the period from 1945 until the end of the 1960s, governments of both a Conservative and a Labour persuasion deviated remarkably little from a policy on State support focused clearly on increasing home-produced food supplies. The policy seemed to be self-evident in its common sense, although in retrospect the key justifications were seen to vary as the years went by. In the immediate aftermath of war, when the memory of the threat of the German submarine menace was fresh and when the views of the majority report of the Scott Committee were still remembered, increased self-sufficiency was an obvious goal. Home production saved food imports which had to be paid for with precious dollars secured from American loans. The introduction of bread rationing in 1946 (something which had not been necessary in wartime) linked to a worldwide shortage of cereals and the need for the British government to

New farm buildings in the countryside. A new farm building typical of type encouraged by the capital investment programmes of the post-war years. The unsympathetic materials (concrete, steel and asbestos), while efficient for farming, provided problems for planners especially in areas of high landscape value such as National Parks, and were increasingly brought within the bounds of planning controls. (By courtesy of Wye College)

find food to counteract near-famine conditions in Germany and India (Marsden *et al.*, 1993), simply confirmed the policy direction. Later on, financial crises which stressed the adverse balance of payments situation also served to justify a policy of support and expansion.

The post-war system, judged by its overall theme of public support for farmers to produce as much as they could, continued well into the 1970s. Britain's eventual joining of the European Community in 1973, after two previous abortive attempts in the 1960s, understandably appeared to mark a watershed, but in practice it was really only the mechanisms of farmer support which were changed. Other European countries, notably France, were if anything even more committed to their farmers who made up a greater proportion of the population than in Britain (Tracy, 1989). National agricultural policy, linked to a series of Agriculture Acts, was replaced by the application of the various elements of the Common Agricultural Policy (CAP). On the major issues of financial support, deficiency payments were replaced by a complex system of intervention levels, target prices, production subsidies and the like funded by the Guarantee Section of the CAP. The broad objective to increase home production remained the same. Thus clear governmental statements to this effect were given in 1975 ('Food from our Own Resources') and again in

1979 ('Farming and the Nation'). While the tone of the latter paper was noticeably less strident than that of its predecessor, it was clearly not intended to change the mind-set of the farmer who had grown up in the post-war world.

The adoption of the CAP naturally brought more than simply the creation of a new regime for financial support. Paralleling the Guarantee element of the Policy (though accounting for no more than 10 per cent of the total budget) was the Guidance Section, committed to policies which were to improve the structure of the farming industry in the Community. Directives from Brussels relating to farm modernization and requiring the creation of farm plans (71/159), early retirement of farmers (72/160), socio-economic advice (72/161) and additional support to farming in 'less favoured areas' (72/268) were taken on board by the British government, though their actual implementation was variable in their relevance and effectiveness. In fact most of these developments, except that on socio-economic advice, had had their precursors in national agricultural policy. The structural directives of the early 1970s were a follow-on from the Mansholt Plan and in some way a curtain-raiser to future policies. In a tentative way they pointed the direction away from the blanket support of farming and farmers simply to produce more food towards much broader rural objectives which tried to relate more directly to questions of overproduction, an ageing farming population and issues of low income particularly among small producers.

In 1988 this move towards a broader approach was consolidated with the attempt to integrate the Guidance Section of the agricultural fund (EAGGF or FEOGA) with the European Regional Fund (ERF) and the European Social Fund (ESF) in a reformulation of the structural measures and their application in line with a set of five main Objectives (Commission of the European Communities, 1989). Of these Objectives, that which is most significant for England and Wales is Objective 5b, relating to 'the promotion of development in rural regions'. In practice, when this policy came into operation in 1989 it only affected parts of rural Wales and of Devon and Cornwall, though the areas granted objective 5b status were subsequently increased in 1993 to include parts of many other counties.

The early 1980s can be seen as the pivot-point for farming policy – not because of any one particular change but because of the coming to a head of several developments which had been in the making for some years. Put simply, the enormous success of the 'productivist' agricultural policies which had been followed since 1945 became increasingly linked with the realization of what this policy was costing the taxpayer. These two elements alone would probably have ensured a very different tone to agricultural policy in the final quarter of the century. But the period also made more public two major ideological developments, each of which would contribute to a radical shift in rural policy – the rise of environmentalism and of the environmental lobby, and the coming to power in Britain (and indeed elsewhere in the developed world) of the New Right, epitomized by the Thatcher administration from 1979. These themes will be returned to later.

The massive state commitment to the farming industry in the years after 1945 was paralleled by significant changes in both the pattern of agricultural land tenure and of farm size. As far as land tenure was concerned, there was a major shift towards owner occupation. In 1950, for example, some 62 per cent of the farmland of England and Wales was rented with just 38 per cent in owner occupation (Ministry of Agriculture, Fisheries and Food, 1968). Forty years on, the proportion had been exactly reversed with two-thirds of agricultural land farmed by owner occupiers (Burrell *et al.*, 1990). The size of farms also increased. At the end of the war just one-quarter of holdings in England and Wales were classified as large, that is having more than 120 hectares of land in crops and grass (Grigg, 1987). By the early 1980s this proportion had doubled and there had been a commensurate reduction in small (under 40 hectares) farms as farm holdings had been amalgamated.

These changes in tenure and size were of course not independent of the economic support given to farmers by the state. Put simply the financial aid to farmers was effectively capitalized into land values and thereby into increasingly high land prices. The incomes of farmers, and especially large farmers, coupled with the willingness of the banks to lend money in large quantities, inevitably encouraged tenants to buy their farms and existing owners to add to their land holdings. This process continued largely uncriticized until the 1980s when its insidious influence was especially attacked by Richard Body (1982) and when the changing fortunes of agriculture forced a re-evaluation of returns which was reflected in a fall in the price of agricultural land.

PLANNING FOR FORESTRY

State support for forestry in the second half of the twentieth century mirrors that given to agriculture – wartime concerns about self-sufficiency, legislation in the immediate post-war period, variations in the level of support over the years and a noticeable change from a simple concern for timber production to wider issues of nature conservation and recreational access.

A White Paper of 1943 (Cmd 6447) stressed the need to increase the home production of timber and set a target for the planting of rather more than 2 million hectares by the end of the century. As with food production, the logic of the policy was founded essentially in strategic concerns. With no more than 5 per cent of the area of England and Wales under forest and woodland and with the need to import around 90 per cent of the country's timber needs, the focus was understandable. The increase would be achieved in two prime ways – continued activity by the Forestry Commission and aid to private growers, both for new planting and improved management practices.

Legislation in 1945 and 1947 created the system which, with some variation, would operate for two decades. Grants for planting, and later an annual grant for management, were made available and a 'dedication scheme' was introduced whereby owners could covenant their land for timber growing over a period of time. As with agriculture, there were times during the 1950s when the cost of these schemes was questioned, though of course they were only ever a small fraction of the monies expended on supporting agriculture. By and large, however, the schemes of grants and management advice were productive such that by the end of the 1960s private woodland owners were planting at the rate of 20,000 hectares each year.

The post-war commitment to increase the forest area was, however, always shadowed by a nagging concern about the low rate of return on forestry. As the memory of wartime shortages became hazier, so the strategic argument weakened. To some extent the economic concerns were countered by the argument that forestry created jobs especially in the hills and uplands and that a social benefit as opposed to a narrowly economic one accrued to these areas. By the early 1970s, however, the economic issues were paramount. A government review of forestry policy (Ministry of Agriculture, Fisheries and Food, 1972) suggested that forestry was delivering economic returns of no more than 3–5 per cent whereas it was conventionally, if somewhat unrealistically, argued that a 10 per cent return was required. While government continued to view forestry as an important investment in job creation in the remote upland areas, the effect of the review was to cut back substantially on the level of State support. The dedication scheme was suspended as was the existing grant scheme. While new schemes were subsequently introduced, the result was a much reduced level of planting throughout the 1970s.

Four main elements characterize forestry policy since the early 1980s. First the political changes following the election of the Thatcher government in 1979 have had an inevitable impact especially upon the activity of the Forestry Commission. Following the 1981 Forestry Act the Commission was directed to adopt a more business-like approach to its operation and in particular to sell off some forest land. Sales continued throughout the 1980s and were widely seen as a build-up to a more thorough going privatization of the agency. By 1994, however, government had retreated from this idea.

Second there were continued changes in the fiscal environment within which forestry operated. Tax concessions to private landowners had always been an important (some would say essential) factor in encouraging tree planting. But in the 1980s concern grew that these concessions, whereby the cost of forestry investments could be set against earnings from other businesses, were being used in a wider context by rich people investing in forestry planting as a convenient and highly profitable means of increasing income. Well-known personalities in particular were publicized as having invested heavily in large-scale coniferous plantations which were also argued as having deleterious effects upon nature conservation and landscape. Even the free-ranging attitude of the Thatcher government to wealth creation had difficulty in accepting this consequence of state forest policy and the major concessions in this area were removed in the 1988 Budget. There was an inevitable reaction to these tax changes seen in the rate of planting especially in the poor lands of the uplands. From over 22,000 hectares in the year 1988/89, the annual rate had dropped to just 7000 hectares by 1992/93.

Third there was a growing concern that forest policy should have broader objectives than simply the increased production of timber. In particular the need to improve the nature conservation, landscape and recreational value of woodlands was increasingly stressed. A broadleaved woodland grant scheme was introduced in 1985 to encourage a greater variety in planting schemes. The need for a greater appreciation of the needs of nature conservation was argued by the Nature Conservancy Council in a report published in 1986 (Nature Conservancy Council, 1986).

The links between forests and recreation, while they have been increasingly stressed in the last twenty years, go back much further. Forest paths had been opened, for example, in Snowdonia in 1937 and in the Forest of Dean a year later (Patmore, 1983) but these were predominantly passive ventures with minimal attempts to cater for or encourage visitors. The growth of the recreational use of the countryside during the 1960s (chapter 8) led to the Forestry Commission being allowed to develop recreation facilities in its forests following the 1967 Forestry Act. Further powers were added by the 1968 Countryside Act. Some developments of holiday accommodation and the like have followed, though not on the scale which was envisaged in the heady days of the recreation boom of the late 1960s.

The fourth strand which can be recognized in policy relates to a growing interest in the potential for forestry and woodland to contribute to the move towards diversification within the rural economy. In particular the 1980s saw an increasing interest in the potential for more woodlands in the lowlands linked in to the normal farm economy. A Farm Woodland Scheme was introduced in 1988 and subsequently updated in 1992 with the aims of diverting land in the lowlands from agriculture, enhancing landscape and wildlife potential, increasing recreational opportunities and creating extra employment and income. The scheme allowed for annual payments for up to forty years to farmers who converted farmland to woodland. In an evaluation of the operation of the scheme in England (Gasson and Hill, 1990) it was noted that a majority of participating farmers quoted the enhancement of landscape and conservation opportunities as their main motivation for joining the scheme.

A number of these policy strands have come together during the late 1980s and 1990s in some revolutionary proposals for forestry development in rural and urban fringe areas. Most notably, following a proposal from the Countryside Commission (Countryside Commission, 1987), a new National Forest covering some 40,000 hectares in the Midlands between Charnwood and Needwood Forests has received Government approval. The objectives of this ambitious plan touch only tangentially upon the traditional policy focus of a need for more home-produced timber. Instead the forest had multiple objectives linking recreation and tourism, landscape enhancement, nature conservation and the need to withdraw land from agriculture. Similar objectives relate to the development of a number of 'community forests' following proposals from the Countryside Commission and the Forestry Commission in 1989. Sited in peri-urban areas, such as east London, schemes of tree planting have been initiated involving local authorities and local community groups with the intention of creating 'a rich mosaic of woods and farms, leisure enterprises, nature areas and open spaces (forming) a well-wooded landscape for wildlife, work, recreation and education on the edges of our towns and cities (Wilkinson, 1992).

THE CHANGING POLITICAL ECONOMY

It was the complaint of many farmers and landowners in the inter-war period that no political party appeared to show real interest in their predicament. That charge can scarcely hold for the post-war period when agriculture has invariably been at the centre of major policy concerns. Moreover the importance given to agriculture by the major political parties has been matched by the significant level of power and influence wielded by farmers and their lobby groups, notably the National Farmers' Union.

This influence derives essentially from the power and prestige which the ownership and control of land gives to those who occupy it (Newby *et al.*, 1978). The farm and forestry lands of England and Wales remain predominantly in private hands, though the statistics of land ownership are notoriously poor. The data collected by the Northfield Committee (Northfield, 1979) give the most accurate outline of the pattern of land ownership. These show that 90 per cent of agricultural land remains in

The National Forest. A view of Beacon Hill on the eastern edge of Charnwood Forest, a major recreational area for the East Midlands, showing the mix of open glades and woodland which is a feature of the new landscape. (By courtesy of the National Forest)

private hands with the bulk of the remainder being owned by the 'traditional institutions' such as local authorities, government departments, public industries and agencies and the Oxbridge colleges.

The close relationship between the National Farmers' Union and government was established very soon after the election of the Labour government in 1945. Part of the folklore of British agricultural policy refers to the high regard in which the new Minister, Tom Williams, was held by the farming community. A senior Ministry civil servant commented at the time:

> . . . the civil service takes an impressive view of the drive of the new Labour Ministers. Our own man in particular has pleased the farmers with his forthright manner and energy. (quoted in Marwick 1976)

This mutual commitment, it has been argued (Self and Storing, 1962; Gardner, 1979) was maintained irrespective of which party was in power from the end of the war until at least the 1980s. There developed a form of 'corporatism' whereby policy was settled cosily between the leaders of the industry and the agricultural ministry. While in no way gainsaying the existence of this corporatist partnership, Flynn (1986) and others have argued that there was less uniformity between the political parties in agricultural policy than is generally assumed. In practice, however, this criticism of the established view should not detract from the basic recognition of the very significant amount of power which has been wielded by the farming lobby in the post-war period.

Until the coming to power of the Thatcher administration, this power nexus remained one of the most powerful in the country. Indeed, although the political influence of the agricultural lobby has been challenged by other power blocs with an interest in the use of the countryside, notably the 'building, digging and dumping' lobbies (Marsden *et al.*, 1993), it still remains remarkably powerful. It is still the only single industry with its own Minister of cabinet rank and is still important at the local level in an albeit weakened local government system (Newby *et al.*, 1978; Newby, 1979/85) where farmers and landowners still contribute a significant proportion of the 'great and the good'.

The consequence of this strong political position maintained by farmers since mid-century was reflected in impressive improvements in levels of income, primarily a by-product of the incentives given to production. Until the mid-1970s farmers' incomes rose along with output to create levels of prosperity for many farmers which contrasted strongly with the 1930s. Since the mid-1970s this picture has been modified to the extent, for example, that real farming income in 1988 was some 25 per cent lower than that of 1974 (Britton, 1990). However this apparent reversal of fortune, much lamented by the agricultural community, hides very significant variations, and many farmers, notably in the dairy sector, have certainly seen their incomes continue to hold up. Moreover income from strictly farming activities is but one element to be considered and there is an increasing proportion of farm households which has other sources of income (Gasson, 1988). In the context of growing levels of economic diversification and pluriactivity (Gasson, Shaw and Winter, 1992) it is clear that farmers have been remarkably successful in maintaining their material prosperity.

The same cannot be said of that other element in the farm work force, the farm workers. Their position within the developing political economy of agriculture has effectively been the reverse of their employers (Armstrong, 1988). Even when trade union power was strong in other industries, the power of the farm workers union was always weak. A dispersed work force made membership recruitment and retention difficult. Moreover, as Newby (1977) showed in his study of East Anglia, the long-standing relationships between master and man at the farm level were frequently based upon deference and accordingly weakened any attempt at strong collective bargaining. Farm workers wage levels remain at a level equivalent to little more than three quarters of the average national manual wage, a relative position which has remained unchanged for decades.

FROM NARROW PRODUCTIVISM TO BROADER COUNTRYSIDE POLICIES

Explicit in the stated objectives of the 1947 Agriculture Act was a set of policy goals relating to stability, self-sufficiency, better incomes and living conditions, low food prices and efficient production. By many standards, the post-war policy measures could be judged a remarkable success in the realization of these goals.

For at least a quarter of a century, stability was the order of the day and even in the uncertain agricultural world of the 1970s and 1980s it could quite reasonably be said 'In some ways it seems misplaced to talk of a farm crisis in Britain? (Cox, Lowe and Winter, 1989*b*). In terms of production and self-sufficiency, the combination of agricultural science, the extension service and financial inducements to farmers to adopt better practice led to dramatic changes. An average yield of wheat in the late 1930s would have been around two tonnes per hectare – barely any improvement since the end of the nineteenth century. By the early 1990s wheat yields, which had risen on average by 2.6 per cent each year between 1952 and 1986, averaged more than 6 tonnes per hectare, while yields of milk at least doubled over the same period (Britton, 1990).

Intimately linked to the post-war policy of increased home food production was a debate on land use and land availability which came to the fore in the mid-1950s. The conventional wisdom not unnaturally believed that every farmland acre was precious and that there should be real concern about losses to urban growth. Farmers and landowners, including many in Parliament, fervently argued this view and they were supported by the new army of planners which after all had been created in large part to protect the countryside from urban incursion. This belief in the fundamental importance of agriculture above all other uses had its origin, of course, in the Majority Report of the Scott Committee. Professor Dennison's Minority Report was largely forgotten. In fact this fundamentalist view remained in place, both officially (where of course it was assiduously cultivated by the National Farmers' Union) and in the public imagination, for decades to come.

Professor Dennison's message was, however, not entirely forgotten. From 1954 a research programme at Wye College, London University, led by Gerald Wibberley, provided much-needed data and a strong challenge to the conventional wisdom (Wibberley, 1959). For the first time the true extent of the urban area and the pace of land loss to the towns was known, thanks to pioneering work by one of Wibberley's colleagues, Robin Best (Best, 1959, 1981). The evidence was clear that the threat of urban growth was being grossly exaggerated. These findings were built on in the 1960s by Wibberley and his co-workers. 'Land budget studies' produced in the mid-1960s (Gasson, 1966*a*) and in 1971 (Edwards and Wibberley, 1971) showed unequivocally that slow population growth coupled with the growth in agricultural productivity would mean sufficient land into the future. Indeed the percentage productivity improvement was about ten times that of the rate of land loss. As Edwards and Wibberley concluded:

> . . . it is possible for home agriculture to feed a growing population and maintain or possibly increase self-sufficiency levels and at the same time release land for urban growth, forestry and recreation at a rate which allows adequately for their likely future development. (page 107)

Viewed from the 1990s, when food surpluses and withdrawal of land from agriculture are commonplace, these studies appear prescient and far sighted. Indeed it could well be argued that they were unduly cautious. For long, however, these views were both unheeded and unpopular. Vested interests, such as farmers and the major countryside protection groups, found the evidence unpalatable. They were supported by a growing preservationist lobby in the country and by some academics and writers who held firm to the agricultural fundamentalism which Wibberley and Best had deplored (Coleman, 1977; Moss, 1978).

The 'land use debate' of the 1950s and 1960s changed its focus and broadened its perspective. The evidence of land loss gradually became accepted in many quarters, not least by planners, and concern shifted towards landscape deterioration, first from urban growth but increasingly from agriculture (Westmacott and Worthington, 1974). The paradox of this change is obvious. Agriculture, which had from the time of the Scott Report been heralded as the protector of the countryside, was now seen as its destroyer.

From the 1970s the criticism extended beyond the question of the destruction of landscapes and increasingly took on a more broad 'environmental' tone. Research which had been carried out by Moore (1969) and others on the harmful effects of agricultural chemicals upon wildlife, widened in the 1970s to become a full-scale assault by conservationists concerned about the environmental damage wrought by farming (Green, 1985). The loss of woodland and meadows, the destruction of hedgerows and the decline of animal and plant species were highlighted in an official report by the Nature Conservancy Council (NCC, 1977). Growing concern about these changes was a major stimulus to the development and passing of the Wildlife and Countryside Act of 1981 (chapter 8). A further study by the Nature Conservancy Council in 1984 re-emphasized the post-war trend, showing, for example that only 3 per cent of lowland meadows which had existed in 1945 remained unaffected by farming and that between a third and a half of all ancient woodlands had disappeared over the same period (NCC, 1984). The loss of hedgerows consequent upon field enlargement linked to machinery use and supposed economics of scale in farming have continued to be an especial focus of concern. As such the loss has become something of a key indicator of the effects of modern agricultural practices on the appearance of the countryside. The evidence in this regard suggests that, despite major changes of attitude in farming, the loss is continuing – between 1984 and 1990 a further 100,000 kilometres of hedgerows were removed in England and Wales (Barr *et al.*, 1991).

Concern in the 1970s for the environmental effects of farming focused also on a number of particular *causes célèbres*. Notable examples were the destruction of moorland by agricultural improvement in the National Parks especially the North York Moors (Parry *et al.*, 1982) and in Exmoor where a special inquiry by Lord Porchester was instituted in 1977. Wetlands, too, were under agricultural pressure as farmers sought to drain land, improve pasture and so increase stocking rates. While wetland reclamation was common in many areas around the coast, it was in the Norfolk Broads and in particular the Halvergate Marshes (Lowe *et al.*, 1986) where the issues became most polarized.

In retrospect three characteristics stand out about the emerging 'farming and environment' debate of the 1970s. The first is that it increasingly concerned a wider public audience than simply informed scientists or conservation specialists. The growth of the environmental lobby (chapter 4; Lowe and Goyder, 1983) ensured that a substantial number of the general public were made aware of the issues and inevitably helped form a new political awareness regarding agricultural policy. This growing public concern was easily fuelled by the writings of such vocal critics as Marion Shoard who, in her polemic *The Theft of the Countryside* (1980), railed against the farmers who, she argued, had at the nation's expense employed the much-vaunted new technologies of the second agricultural revolution to destroy the countryside. The argument as to how the powerful farming lobby had effectively alienated three-quarters of the land area of the country for their exclusive use was further elaborated in her second critical attack *This Land is Our Land* (1987).

A second characteristic is that the emerging debate was both fragmented and was often focused on quite particular and narrow conservation issues. Particular pressure groups had their especial concerns, perhaps about the loss of certain wildlife species. Particular locations gave rise to intense publicity because they were seen as being exceptional. Only gradually over the decade did there develop an overall concern about the countryside as a whole and about the perceived threat to ordinary landscapes rather than especially precious ones.

A third characteristic was that the concentration on particular conservation concerns, often in particular places, tended to hold back a realization that it was the agricultural policies which had been set up after 1947 and which were continued by the EC Common Agricultural Policy which were the ultimate engine of change. An important analysis by the economists Bowers and Cheshire (1983) made the point clear. It was not so much the wilful and ignorant actions of farmers which were consciously destroying the countryside, but rather an economic policy which provided rewards for increased production but little material recompense for 'environmentally friendly' farming systems (Potter, 1986). Moreover, as was seen earlier, farmers were still encouraged by the Ministry of Agriculture during the 1970s to produce more food from home resources. Little wonder then that most farmers responded to political exhortation and market signals in the way that they did.

The 1980s saw a radical change from the post-war productionist policies which had run for thirty years or more. The change was brought about by a coming together of a number of strands, chief of which were the realization of the enormous cost of the support system which once had seemed so necessary, coupled with the mounting stocks of foodstuffs in the European Community (Cox, Lowe and Winter, 1989*a*). On the home front the enormity of the cost of farm support was trenchantly argued by the Tory backbencher Richard Body in a brace of polemical books published in the early 1980s (Body, 1982, 1984). While Body was especially outspoken in his views, they were not too far out of line with the emerging political consensus ushered in by Margaret Thatcher's 1979 election victory. In particular the Conservative government was keen to see an end of the old liaison between the farmers and government which had centred upon the annual price review and the establishment of a price regime much more in line with market forces.

A further strand was, of course, the continued criticism of the environmental damage caused by agriculture. The new situation of the growing importance of the environmental vote helped change the composition of the groups which together set the rural policy agenda.

Gradually, and in a sometimes incoherent way as the decade progressed, these three strands of mounting economic cost, changing political ideology and environmental concern came together. Attempts to control production centred first (from 1984) upon the imposition within the European Community of quotas to control milk production. In 1988 a voluntary 'set-aside' scheme was introduced whereby participating farmers were required to withdraw 20 per cent of their land from arable production (Potter *et al.*, 1991). By the end of the decade proposals from the EC Agricultural Commissioner Ray McSharry had introduced reduced commodity prices for cereals, milk and beef. As was seen earlier in this chapter, the changed political climate of the 1980s led to an undoubted waning in the lobby power of the farmers' unions which had been a major prop to the old corporatist political economy. And finally the 1980s saw the introduction of more broad-based environmental policies which, albeit with some difficulty, began to be linked with payments for adoption and therefore to reward farmers for farming in environmentally sensitive ways and not just for producing food.

The furore which developed around the draining of the Halvergate Marshes had resulted in a pilot scheme whereby farmers were paid for keeping their land in its wetland state. This led directly to the creation in the Agriculture Act 1986 of a scheme for Environmentally Sensitive Areas (ESA) and which was incorporated into the Structures Regulation of the European Community. Originally some six such areas, chosen from an original list of 46 suggested sites, were defined where farmers were to be paid for farming in ways which were seen as having created valuable landscapes and wildlife habitats. By the early 1990s there were 29 ESAs in England and Wales and further proposals made in 1993 would increase the area of land in England thus designated to 1.17 million hectares.

Linking these specific attempts to encourage farmers to use their land in environmentally suitable ways, came a broad package of policies introduced in 1987 under the title of ALURE – alternative land use and rural economy (Cloke and McLoughlin, 1989). While part of these proposals related specifically to environmental concerns – notably an increase in the number of ESA's – the broad thrust of the package was intended to help farmers diversify their activities from food production towards other uses for land. These included schemes to encourage the planting of woodlands on farms and to expand the private forestry sector and a diversification scheme to encourage farm businesses in such areas as tourism and small industry.

Proposals made in 1993 involving an increase in the number of ESA's, an expansion of the scheme to protect 'nitrate sensitive areas' and a proposal to combat the overgrazing of moorland meant that, within a relatively short period, something approaching a coherent set of agri-environment policies were in force (Potter, 1993). The package provided a clear indication of the major changes which had taken place in the previous forty years. A state-centred farming policy, unequivocally focused upon food production and automatically equated with countryside care, had become a more market-based system linked to a broader-based rural economy in which so-called 'CARE' (countryside, amenity, recreation, environment) goods are specifically rewarded.

CONCLUSION

It was noted in the previous chapter that the majority membership of the Scott Committee were convinced, not only that the revival of traditional mixed farming would be a necessary element in the future health of the countryside, but also that there would be no 'radical alteration' in either types of farming or the resulting pattern of landscape in the open countryside.

Half a century after the publication of this view, it is difficult to envisage a forecast which could be more wide of the mark. Mixed farming, of course, is still to be found, but the overall trend on farms has been towards specialization. In particular, there has been a polarization towards arable cropping in the east of England and towards livestock systems in Wales and the west of England. Within traditional crop and animal enterprises the techniques and scale of operations have also changed radically. For example, large-scale units in poultry production, involving flocks of several thousands and in some cases hundreds of thousands, were unheard of in the 1940s but by the mid-1960s were beginning to become commonplace. Equivalent changes have been evident in cereal production where fields have been enlarged to accommodate larger machinery with the consequent loss of hedgerows and field boundaries noted earlier.

In recent years the changes in the agricultural landscape have appeared especially dramatic. As incomes from conventional farming activities have faltered, so farmers have turned to new and sometimes novel enterprises. Thus farm buildings have been converted for holiday accommodation while former cropland is now used for golf courses. Perhaps the most evident signs of change have related to the introduction of crops such as oilseed rape, linseed and lupin which, with their vibrant colours, have added a startling dimension to the lowland landscape.

The paradox, of course, is that many of these changes, which would surely have horrified the members of Lord Justice Scott's Committee, have been brought about by circumstances and policy mechanisms which in large part they would have applauded and which either explicitly or implicitly were at the heart of their recommendations. State support for farming activities and the protective umbrella of the planning system allowed farmers free rein to respond to economic signals. The combination of a free extension service coupled with continued exhortation by government to produce more food by implementing modern scientific methods, meant that farmers were always keen to take on board every new technological development. The mistake, of course, was to view farming as an inherently benign activity and one which, as far as the broad health of the countryside was concerned, should be seen as the trusted ally and linchpin of protective rural planning. In the event the rural changes which resulted were very different from the planning outcomes that were expected.

7

RURAL ECONOMIC CHANGE AND POLICY SINCE 1945

The determination by policy makers in the early years after 1945 to build on the commitment to agriculture which had been made in wartime was understandably seen also as a commitment to a broader rural economy. The general view followed the tenor of the Scott Report: a healthy farming industry was a *sine qua non* not only for national food policy and for proper landscape protection, but was also essential for a revival of rural economic fortunes. Agriculture was seen, not just as the major employer in the countryside, but also as the focus for economic activity, with rural industry and crafts providing the necessary service and support. Even as perceptive a commentator as C.S. Orwin clearly viewed the revival of farming as the central plank for rural reconstruction (Orwin, 1945).

There was equally no doubt as to the existence of a distinct *rural* economy which was seen as separate from urban structures both in its character and in its independence. The character was predominantly farming-focused, though some rural areas were clearly important for other primary production such as fishing, mining or quarrying. Where manufacturing existed, it was generally as a service function to primary production such as agricultural machinery or food processing.

The signs that economic change was on the way for the countryside were certainly there in 1945 but at the time they were treated either as an aberration or as a development to be viewed with circumspection. In the 1930s many market towns, particularly in the more prosperous south, had begun to see the growth of some of the 'new' industries based upon modern manufacturing processes such as automobile and aircraft manufacture and electrical engineering. Particularly where the growing trunk road system allowed easy access to markets, new industries had begun to develop which had no specifically rural, and certainly not agricultural, roots. A classic case was that of Banbury in north Oxfordshire. Traditionally the centre of an agricultural region, with its associated industries, it had begun to attract manufacturing industry (related, for example, to aluminium alloy manufacture) from the early 1930s and the character of what had always been a local rural market centre was beginning to change (Stacey, 1960).

In practice, the first two decades after the war saw a form of double-think when it came to the economic development of the countryside. On the one hand the conventional wisdom that agriculture was at the heart of the rural economy was maintained. As the previous chapter showed, the political and financial support given to the farming industry left little doubt about that. On the other hand, the process of encouraging new types of employment in the rural market towns went on apace and was consciously encouraged by regional policy which sought to move firms from crowded urban locations to more spacious rural sites. The 1950s, and particularly the 1960s when the national economy boomed, saw many new manufacturing firms open up in small towns and take on workers no longer needed within agriculture and also increasingly to provide employment for rural women.

This view, of an essentially agricultural economy in the countryside but one which was changing quite rapidly as new manufacturing firms moved in, held sway until probably the end of the 1970s. The dual objectives of substantial support for the domestic farming industry and the diversification of the employment base by the encouragement of small-scale manufacturing industry were parallel but separate policy objectives. Both were seen as legitimate areas for state involvement. The commitment to farming was undoubted and was considered in detail in chapter 6. The commitment to industrial development, particularly in regions which were judged as lagging or deprived, was hardly less enthusiastic, though the prosecution of regional policy naturally waxed and waned as political and economic fortunes changed (Wannop and Cherry, 1994). Inducements to firms by way, for example, of grants, loans and the provision of sites on new industrial estates on the edge of market towns were commonplace. Just as there was seen to be a long-standing rural population problem, so also was there an apparently clear rural employment problem which state aid and the new industries were intended to address.

The period since the end of the 1970s has seen major reassessments in both the understanding of the economy of rural areas and of the role of the State with regard to the development of that economy. In part at least this is due simply to the major political and ideological changes which were heralded by the election of the Conservative government under Margaret Thatcher in 1979. Very quickly many of the old tenets of a distinctly *dirigiste* economic policy were removed. As chapter 6 showed, the favouritism shown to agriculture was also directly questioned. Secondly the role of the State in encouraging economic development in rural areas was limited as regional policy and its associated incentives were drastically cut back and as local authorities found their powers and their financial resources reduced. Thirdly the restrictions on economic development which had been inherent in the post-war land use planning system were eased somewhat in the expectation that private enterprise would provide the new foundation for rural employment and economic growth in the countryside.

These ideological changes were matched by a resurgence of interest in the nature of employment and economic structure in the rural areas. It is not too much of an exaggeration to say that, before the 1980s, it was generally only agricultural economists who were interested in these topics. Few other experts were interested in the 'rural economy' beyond perhaps some regional economists who saw the rural areas essentially as the hinterlands of urban growth poles. Since the 1980s, the rural economy, or perhaps more accurately the economies of rural areas, has been a clear focus of interest and research. Partly this is due to a general resurgence of interest in matters rural, a theme echoed elsewhere in this book. But in the main part it is because the economic structures of rural areas are no longer seen as simply residual elements of agriculture or the spill-over of urban manufacturing. While both of these elements can still be found, the economies of rural areas are now seen as being increasingly characterized by modern industrial structures and processes, often with leading-edge technology, with levels of growth which are well above the national let alone the urban average, and with national and international linkages rather than narrowly local connections.

RURAL EMPLOYMENT AND RESTRUCTURING

Beyond the peripheries of the major conurbations and towns, the economy of the countryside in the late 1940s was still evidently linked to agriculture. While direct employment in farming was low, agriculture was still the indirect basis of much local employment. A half century on, this is demonstrably no longer the case. Certainly in lowland England, and to a significant extent elsewhere, it is increasingly difficult to recognize a distinctively 'rural' economy. Research by Errington (1990), Townsend (1991) and others has shown that the employment patterns to be found in the countryside are remarkably similar to both the overall national pattern and to many urban areas. The dominant proportion, perhaps 60–70 per cent, is to be found in the service industries such as distribution, tourism, transport and banking and finance. Around one-fifth to one-quarter are employed in the manufacturing industries, only a minority of which would be directly linked to food and drink manufacture. No more than 5 per cent of employees remain in agriculture, though this figure would reach perhaps one in ten in the more rural and remote areas. Naturally there are regional variations, with particular employment sectors or trades being important in particular areas (Hodge and Monk, 1991), but the regional individuality of employment is very much less pronounced than was the case in the past. Equally rural connotations of employment may still be evident, though such rural connections may well be more linked to marketing strategy than to any real historical tradition. Thus the traditional tie between food manufacture or brewing as 'rural' industries may reflect historic patterns but the companies concerned will often be national or even multi-national in organization, with few if any ties to the local farming area.

In the second half of the twentieth century rural areas have also exhibited other characteristics in addition to employment types which link them with the wider national economy. Female employment has tended to grow, increasing by around 10 per cent in the rural areas for example in the 1980s (Townsend, 1991), while male employment has decreased. Part-time employment has tended to increase at the expense of full-time jobs, especially in enterprises linked to tourism and recreation. Self-employment, which was always more important in rural areas than in towns, has also grown disproportionately. And finally there has been a notable trend towards the creation and growth of small businesses (North and Smallbone, 1993), a trend consciously fostered by governments which have seen them as a major growth force within the economy as a whole.

It is now clear that, far from being economic backwaters, rural areas have in fact seen major economic growth since mid-century. Just as the population turnaround to counterurbanization was belatedly recognized in the 1980s (chapters 9 and 10), so too can it be argued that a form of 'counterindustrialization' has also been at work and probably for as many decades. The work of Fothergill and Gudgin in the early 1980s (1982) was amongst the first to emphasize this point. They showed unequivocally that the rural areas, far from being 'the places left behind', were in fact often at the centre of economic change and growth. While the manufacturing sector has *nationally* been declining relatively in the post-war period, it has been growing in many rural areas. Thus between 1960 and 1987, while London and the other conurbations lost over two million manufacturing jobs (a relative decrease of over 55 per cent), the rural areas saw a growth of over 100,000 jobs (19.7 per cent increase) (Gudgin, 1990).

The reasons behind the growth in manufacturing industry based in rural areas are many and varied. In large part, however, it is because many companies are engaged in production activities which have few strong locational limitations, and can thus be located virtually anywhere, given that access to markets is relatively easy. Most obviously this can be seen in the 'high-tech' manufacturing sector such as electronics where clean environments, small but highly skilled workforces and few if any bulky raw materials are involved. But in fact those industries are simply the more extreme example of a general characteristic of much wider manufacturing where the key is a relatively high 'value added' end product. In some cases these new enterprises are linked to major national and international corporations, while other enterprises are free-standing and may relate to new processes and manufactures in, for example, telecommunications or the defence industry.

Beyond the lessening influence of traditional locational constraints, the reasons behind the relative success of the manufacturing sector in rural areas are much debated. It has been argued (Fothergill and Gudgin, 1982) that a major factor has been the limitations of space for expansion which successful companies found in their original urban location. Others have pointed to the relatively low wages characteristic of rural areas as an attracting force for entrepreneurs. The absence of strong trade union power in rural areas, which may be seen as inhibiting new developments, may also be important. The general attractiveness of the countryside as a place in which to live and work may also be a key factor for some firms.

While the manufacturing sector has undoubtedly grown in many rural areas since mid-century, it must be remembered that this growth has often been relative to a very small manufacturing base of employment beforehand and has often been at the expense of the urban areas from which industries have migrated. Overall at the national level the manufacturing sector has lost out in the latter part of the twentieth century as the service industries have grown.

Many new manufacturing firms have, of course, themselves encouraged the growth of the service industries which support them, and in addition the countryside has seen the growth of much service employment related to the tourist industry (Johnstone *et al.*, 1990). Thus between 1981 and 1987 the 'remoter, mainly rural' areas of Britain saw an increase of over 22,000 new jobs in tourism-related services, an increase during the period of over 17 per cent (Townsend, 1991). Other sectors of the service industries have also favoured the rural areas, particularly those linked to the defence industries and the armed services which are often important in the rural areas of parts of the south and the east of England.

These major shifts in economic structure need to be seen not as simply a response to a declining agriculture nor even as a simple rural extension to the industrialization of the metropolitan periphery. Rather are they symptomatic of a major shift in broad economic structures worldwide. Within the industrialized countries the significance of the manufacturing sector has changed, both in terms of the type of manufacturing and also its size relative to the economy as a whole. Equally the service industries have come to dominate national economic structures. Rural areas have been at the heart of both these broad trends. And some rural areas have taken the lead in the creation of this new post-industrial economy. Most obviously these areas have been in that great swathe of attractive countryside which stretches from the west of England, especially around Bristol, eastwards to the Thames corridor and Oxford and thence north eastwards to Cambridge and rural Norfolk. This area, termed the Golden Belt by Peter Hall (Blunden and Curry, 1988), epitomizes those aspects of the rural areas which have proved attractive to entrepreneurs and the growth industries. They provide locations that are accessible to national and international markets and to the world of business and they provide an environment which is attractive for both work and residence. Moreover these areas, especially in the south east of England, have, it has been argued (Fielding, 1992), provided an 'engine' for the economy as a whole. These southern rural areas have provided the start for new enterprises and for a new managerial class which have then spread to the rest of the country, taking with them the economic and associated social changes to 'colonize' the countrysides of the north and west.

What remains of the 'traditional' rural economy based essentially upon primary industry? Beyond agriculture, the other traditional employers in this sector have largely disappeared from rural areas. Quarrying and mineral extraction remain significant in some areas, although the levels of employment are low in an industry where mechanization has come to dominate. More obviously the virtual elimination of the coal industry has totally destroyed the local primary economies of such areas as West Durham (Hubbard, 1982), rural Nottinghamshire or East Kent.

The changes in agricultural employment have scarcely been less dramatic, although the continued dominance of farming in the rural landscape inevitably masks the changes. As far as farmers were concerned, there were just about as many operating in the inter-war period as in the middle of the nineteenth century. A decline after the Second World War, linked to the amalgamation of farm holdings and the growing size of farm businesses, meant that by the early 1990s there were just 136,000 full time farmers in England and Wales (agricultural statistics, Ministry of Agriculture, Fisheries and Food).

Much greater changes have occurred in the numbers employed as hired non-family workers which have in fact been declining throughout the whole period covered by this book. A temporary increase in farm employment during and just after the Second World War did nothing to halt the decline. Some 554,000 were employed full time on farms in England and Wales in 1951 but just 77,000 remained by 1992. In practice, as Clark (1991) and others argue, this decline is a little misleading and should not be seen as simply a 'drift from the land'. In fact the sector has in its own way been quite dynamic, for while it has recorded a net loss of around 10,000 people a year, this has been made up of a continued recruitment of young workers who then tend to leave for better prospects after a few years. Farms with non-family hired workers are now in a minority and the agricultural crisis of the 1980s and 1990s has made the dependence on family labour even more pronounced (Gasson and Errington, 1993).

At least as significant as the broad changes of numbers employed in farming have been the changes in the characteristics of the workers and in the nature of work done on farms. Agriculture has shared in the major structural changes exhibited by other sectors of the economy. Thus part-time employment has tended to take over from full-time work and farm tasks are commonly carried out either by outside contractors or by seasonally-employed casual labour (Errington, 1988; Gasson, 1988). Equally the diversification of agricultural businesses (chapter 6) has meant that employees, and particularly family workers, may be found engaged in a variety of occupations linked, for example, to tourism enterprises, retail outlets and conservation work (Gasson and Errington, 1993).

This diversification into 'non-traditional' forms of work on farm holdings has been partly responsible for focusing some attention on women as a major, if under-reported, element of the farm work force. While the farmer's wife (and probably his daughters) were somewhat condescendingly assumed to support the farm economy, the nature of that contribution has become clearer in recent years. Thus Gasson (1992) has recognized the varying roles of women on farms as worksharers, housewives and, sometimes, equal farming partners, while other workers (e.g. Little, 1987; Whatmore, 1991) have been concerned to explore these roles within wider aspects of social relations.

The underlying theme behind changes in farm-based employment during this period can be summed up in the term 'pluriactivity'. The farm business of today has several strands and they are not necessarily linked directly to the production of food and fibre. For many farmers this change from a single-focus to a multi-focus has been a painful and unpopular process, forced by economic necessity. For some the change is too much and they have left the task to others or the next generation. It should not be thought, however, that such economic diversification has its origins only in the farm crisis of the 1980s and 1990s. In what is now rightly regarded as a classic investigation, Gasson (1966*b*) showed how new patterns of land ownership in Kent and Surrey, linked to growing urbanization and the proximity of London, were already leading to new forms of land use and economic enterprise. The patterns of incipient pluriactivity which she identified in the early 1960s have in the subsequent thirty years spread to virtually the whole of the countryside.

The other element in the 'traditional' rural economy, rural crafts and trades, had also gone through a period of change since mid-century. By the outbreak of the Second World War the numbers of blacksmiths, carpenters, thatchers, saddlers and the like had substantially declined.

John Saville's study of population change in the South Hams of Devon showed a halving in the number of such craftsmen from 118 in 1910 to just 61 in 1939 (Saville, 1957) and similar losses were recorded for other rural areas (Chalklin, 1989). The decline continued throughout the 1950s and 1960s. Of course the original *raison d'être* of such crafts lay in their links as service trades to the farming industry. As agriculture modernized, so the demand for craft products fell. There were fewer horses needing farriers and saddlers and fewer farm buildings needing local bricks and tiles when asbestos and cement were cheaper. Moreover the wages in the quarries of Stonesfield producing the characteristic Cotswold 'slates' or roof tiles could easily be bettered by work in the factories of nearby Banbury or Oxford.

The last quarter of the century, however, has seen a distinct revival in many of these crafts and trades but the growth in demand for their products are in no way related to the agricultural economy. Rather are they an economic by-product of the demographic and social changes and rural affluence which are outlined in chapter 9. Thus farriers and saddlers

Changing country crafts. A blacksmith at work in the early 1960s making a wrought iron gate. A change from the farm implement repairs and farriery characteristic of the development of country crafts in the post-war period. (By courtesy of the Rural History Centre, University of Reading)

Table 7.1. Employment change in Devon by sector, 1951–1991.

Sector	*1951* *per cent*	*1981* *per cent*	*1991* *per cent*
Agriculture, forestry and farming	22.71	5.33	4.51
Mining and quarrying	1.55	0.80	1.75
Energy and water	1.39	1.62	1.37
Manufacturing	12.88	17.77	13.73
Construction	8.37	7.50	8.23
Distribution and catering	12.41	15.36	23.72
Communications and transport	6.03	5.03	5.62
Insurance, banking and finance	1.36	2.86	9.42
Other services	33.31	43.74	31.65
Total	100.00	100.00	100.00

Source: General Register Office, Census 1951, Industry Tables; OPCS Census 1981, Economic Activity Tables; Boucher, Flynn and Lowe (1991). OPCS Census 1991, Economic Activity Tables.

find increasing work in many rural areas, not with farm horses but with the horses and ponies of the middle classes, while the thatcher and the skilled blacksmith are in demand to beautify the houses of rural newcomers.

This account of employment change since 1945 may be concluded by a brief reference to the county of Devon which has been used by Boucher *et al.* (1991) as the focus of a study on rural employment change and policy. Table 7.1 shows the broad structure of employment in this rural county in 1951, 1981 and 1991. The massive decline in agricultural employment is very evident as is the growth in distribution and catering services related to a burgeoning tourism industry. In common with many rural areas, the level of self employment is higher than the national average, as is employment in small firms rather than large enterprises. Of particular note during the 1980s was the very significant growth in financial services, indicating how growth in this particular sector, which was characteristic nationally of the 1980s, was by no means restricted to the major urban financial centres. Boucher's study is valuable, not just because it records broad structural changes in the economy of a rural county, but because she is able to detail the complex network of state planning as it affects rural economies. This network provides the focus for the next section.

PLANNING FOR RURAL DEVELOPMENT

It will be clear from what has been said so far that the course of rural economic change since 1945 has generally not been one of the stagnation of traditional industries and the marginalization of local economies. Moreover, if many of the major developments in economic growth can be seen as being linked to broader patterns of change, it begs the question as to the role of state influence and support in creating or contributing to economic growth and success. Whatever the reader's ideological predispositions, there is a powerful case to argue that the bulk of rural growth has come about because of national and international changes in the economies of the rich nations and has only marginally been influenced by state intervention and formal rural development policy.

Nevertheless, whatever the truth of this view, it is important to recognize the role which government in various guises has had in influencing rural economic change since 1945. In outline it may be argued that there have been four major areas in which state policy to influence rural economic structures have been significant. These have been:

(*i*) support to the agricultural industry;
(*ii*) decentralization of employment from crowded urban areas;
(*iii*) programmes of regional planning and aid to designated backward areas;
(*iv*) the role of the rural development agencies.

The first of these has been dealt with substantially in the last chapter and need not be considered much further here. It should be recognized, however, just how significant in overall financial terms is this aid to agriculture when compared with other types of support for rural areas. In a pioneering study to gauge the level of support for rural areas, Hill and Young (1991) estimated that, while there were no fewer than 170 individual programmes delivering aid to rural areas, no less than 80 per cent of all monies went to support farmers' incomes. From the late 1980s there has been a very slight change in the nature of this support as some payments to farmers have been redirected towards non-food producing ends. This is perhaps most clearly evident in the area of tourism and recreation development; thus between 1988 and 1990 some 60 per cent of capital grants payable under the Farm Diversification Grant Scheme in England were for tourism and recreation enterprises (Boucher, Flynn and Lowe, 1991).

A major element of development policy in the years after the end of the war related to controlling the growth of large towns and cities (chapter 10). As part of this overall strategy, many country towns were seen as suitable recipients of urban overspill population and of industries which should not be allowed to expand in the major urban areas. The classic instance of this approach focused upon the Town Development Act of 1952 whereby small country towns entered into agreement with London County Council (and later the Greater London Council), Birmingham and other major urban centres to accept both people and jobs decanted from those centres. The industries which grew up from the 1950s encouraged by this policy, together with the new firms which were attracted in their own right to the new industrial estates on the edge of these growth points, revolutionized the employment structures of the country towns and their hinterlands. A classic case in point was Ashford in Kent. An agricultural market town of long standing, it had developed as a railway centre in the nineteenth century. Its designation as a growth centre and the acceptance of a substantial number of new people and associated industrial development as London overspill, however, totally changed its economic structure and that of its surrounding rural area. The agricultural and railway functions remained as residual activities, but they were overwhelmed by new employment in plastics, engineering, food processing and chemicals. This pattern was repeated, particularly in the southern half of England where modest agricultural market towns became minor industrial centres in the post-war years, each with its grouping of industrial estates on the urban periphery.

The expansion of market towns – Watercress Farm estate, Ashford, Kent. Typical development on the edges of small market towns linked to overspill population and the development of new industries following the Town Development Act of 1952. (By courtesy of Ashford Borough Council)

This growth of employment was of course a process which was mediated by the land use planning system – indeed the growth of country towns was simply a reciprocal effect of the attempt to control urban growth. Elsewhere in the countryside the town and country planning system continued its traditional role as a protector of land resources and landscapes, generally refusing industrial pressures other than those which were focused upon the small towns. By the early 1980s, however, this role was beginning to change. The election of the Thatcher government brought with it a radical questioning of the control exercised by the planning system coupled with a desire to decrease the level of national unemployment. Thus the tone regarding attitudes to rural development demonstrably changed. Policy advice in 1980 (Department of the Environment, 1980) began to encourage local planning authorities to be more positive in giving planning permission for developments which might create rural jobs, followed by more comprehensive instruction by way of ministerial exhortation and planning policy guidance (Department of the Environment 1992*c*).

The third main vehicle for planned rural economic change has been essentially a product of the regional planning system and of the broader role of local authorities in economic development. The origins of regional planning in the 1930s were of course closely linked to the deteriorating conditions to be found in the major urban areas and were especially tied to the decline of the nineteenth-century industrial activities of those centres. But from the start rural areas inevitably figured in the pattern of regional planning simply by being the hinterlands of the problem urban centres. Moreover some areas which were formally defined as Development Areas were more rural than those centred upon the big conurbations, for example West Durham or the Cumbria coast around Whitehaven and Millom.

The boundaries of the various 'problem area' designations varied from time to time as legislation changed and particularly in the 1960s and 1970s as the proportion of the country covered by these designations expanded substantially. By the end of the 1970s substantial areas of the north of England and virtually the whole of Wales, as well as parts of south-west England were covered by some regional planning designation. The extent, however, to which this state action actually benefited the rural areas is uncertain. In the first place, the broad economic changes which were highlighted earlier in this chapter must inevitably raise the question as to whether rural employment growth where it did occur would not have happened anyway. Secondly there is a strong suggestion that the real focus of benefit was concentrated upon the 'growth pole' with little employment and investment 'trickling down' to the broader rural hinterland. Thus Plymouth gained substantially from state aid, and from its links with the defence industry, but there is little evidence that this was connected to any economic growth which occurred in the market towns of Devon. In a similar way, though outwith the formal regional planning designation, Moseley (1973, 1974) showed how investment and industrial growth in the larger urban centres of East Anglia gave little benefit to the small centres and villages by way of employment creation, or raw material resourcing.

The final, and perhaps most obvious, basis for planned economic change in the post-war countryside has been the activities of the various rural development agencies. At mid-century this really only meant the Development Commission, founded in 1909, and with a general responsibility over the whole of Great Britain to combat rural depopulation by programmes of economic and social regeneration. For much of the early part of the period under consideration, the Development Commission continued to operate in the traditional way which it had built up in the inter-war period, focusing especially upon supporting country crafts through the medium of its own Rural Industries Bureau (Clarkson, 1980). These crafts were, as noted above, still seen as essential rural industries and as necessary adjuncts to the maintenance of agriculture. The activity level of the Development Commission in the 1950s and 1960s appears in retrospect to have been leisurely and low-key; the Commission itself met under the relaxed chairmanship of Lady Albemarle and included among its members the redoubtable Leonard Elmhirst, the founder of Dartington Hall (Young, 1982). Reports of the work of the Commission were infrequent, indeed they had been suspended during the Second World War and were not resumed until 1961.

Until the mid-1960s the work of the Development Commission was rarely proactive but was rather at the response of particular local authorities and generally consisted of making loans from the Development Fund. Minay (1985, 1990) has argued that the approach was largely *ad hoc* and hardly amounted to any form of rural development strategy. This situation began to change in the mid-1960s and particularly so after 1974 under the dynamic leadership of a new chairman, Donald Chapman, later Lord Northfield. Starting in mid-Wales, advance factories were built under the auspices of the Mid-Wales Industrial Development Association which had been created in 1957. This approach was extended from 1965 when the Commission began to recognize so-called ‘trigger areas’ which were based on growth points in problem rural areas. A strategy thus began gradually to develop and programmes of rural development became more innovative and adventurous. In 1973 the trigger areas, which had in effect been an experiment towards the development of a strategy, were replaced by a more comprehensive system of Special Investment Areas and by a rolling programme of building advance factory units. Local authorities were required to submit Action Plans indicating particular needs and problems and outlining the action required.

In 1976 the work of the Development Commission was refocused following the publication of an interdepartmental study which had identified the main areas still suffering from rural depopulation (HM Treasury, 1976). With hindsight this restatement of population loss as the essential framework for identifying rural problems appears curious. The 1971 Census had begun to indicate that many rural areas were in fact gaining population and indeed the Development Commission itself had commented on this phenomenon in its reports. Bearing in mind what is now understood about the very imperfect (arguably non-existent) relationships between population loss, economic growth and employment levels, and about the growth of the counterurbanization phenomenon, this return to the tenets of the original 1909 focus seems bizarre. It was not until 1984 that this simplistic link was changed and other factors such as unemployment levels and service losses were taken into consideration in designating areas for special policies.

The 1960s saw major changes in the area of jurisdiction over which the Commission had responsibility. General responsibilities in Scotland were devolved to the Scottish Office and more specifically within the area of the crofting counties to the Highlands and Islands Development Board which was created in 1965 (Grassie, 1983). An equivalent change took place at about the same time with the designation of mid-Wales as a Development Area in 1966 and culminating in the creation of the Development Board for Rural Wales (now Mid-Wales Development) in 1977 (Lewis, 1991). By then the Development Commission was calling itself 'England's rural development agency'.

The 1960s also saw major changes in some other aspects of the structure and operation of the Development Commission. The old Rural Industries Bureau, which had tended to concentrate on the traditional crafts, was replaced in 1968 by the Council for Small Industries in Rural Areas (CoSIRA), with a much more pro-active brief to encourage 'modern' industry in rural areas.

This programme of problem area designation has continued, with the replacement of the Special Investment Areas in 1984 with a system of Rural Development Areas (Bowler and Lewis, 1991) covering large parts of the rural periphery of England. The Rural Development Areas were re-designated in 1993 (Rural Development Commission, 1993) and now cover some 35 per cent of England and 6 per cent of the population. There is little doubt that the operation of the agency (renamed the *Rural* Development Commission in 1988 when CoSIRA was merged back into the parent body) has, since the early 1970s, become more professional in both its approach and its level of operations. In the early 1990s it operated a budget in excess of £40 million with a staff of around 350.

The 1960s also saw a short-lived, but significant, development involving

Technological support for rural businesses and communities in the 1990s. The Romney Resource Centre acts as a 'telecottage' for small businesses and local communities in the Kent Rural Development Area. Funded by the Rural Development Commission, Kent County Council and private sponsors, the Centre provides modern technology and telecommunications to overcome problems of isolation. (By courtesy of Kent County Council)

another agency of rural development. Under the provisions of the Agriculture Act 1967, there could be designated so-called 'rural development boards' in upland areas where rural problems were seen as especially severe. Such boards would have powers, not only to encourage rural development and enterprise generally, but which were more significantly focused upon the need to improve agricultural structures. Thus they were given powers to intervene in the land market, for example to prevent undesirable farm amalgamation.

In practice only one such board, the North Pennines Rural Development Board, was created, although a second board was planned for mid-Wales. The Board had already begun to use its powers regarding land sales (House, 1976) when the election of a Conservative government in 1970 led swiftly to its demise. The Board's activities, especially in the land market and in forestry, had already resulted in antagonism from local landowners and the national political changes brought its operation into direct ideological conflict.

In Wales, the activities of agencies have been predicated upon a 'top down' industrial development policy, classically centred upon urban growth points. Following the creation in 1957 of the Mid-Wales Industrial Development Association (Gabett-Edwards, 1972), 1967 saw the creation of a Development Corporation based upon Newtown with a brief to attract industry and population to mid-Wales. This approach has been continued by the Development Board for Rural Wales from 1977 such that over 500 workshops and factories, many of them branch plants of major national or international companies, had been created by the 1990s (Lewis, 1991). The activities of rural development agencies in Wales, and especially those of Mid-Wales Development, have been frequently heralded as a success story, brought about by strong state action following years of internecine local political struggle and reflected in numbers of jobs created and, more debateably, in an undeniable growth in the population, at least of the main settlements. More recently the apparent success has been criticized as superficial. Thus Day *et al.* (1989) have, based upon work in Wales, pointed to the evidence of firms cutting down on labour once they have established, and to the reality of many poor quality jobs behind the apparent numerical success. Too many jobs, it is argued, are poorly-paid, lack training and prospects and are directed disproportionately to part-time and women workers. It should be noted, perhaps, that this is simply a particular case of much apparently successful rural development, whether initiated by state agencies or coming from independent entrepreneurial action.

A EUROPEAN DIMENSION

The major directions of rural economic development policy which were outlined above were essentially national, that is they devolved from planning structures vested in national legislation and national agencies. The last decade or so, however, has seen the consolidation of another element in rural planning: the involvement of a wider European perspective.

As this chapter argued at the outset, national policy towards the economic wellbeing of the countryside was for a long time founded in the support of the agricultural economy. Only in more recent years have there been very significant divergences from this focus. Much the same applies to the European approach to rural policy. For most of the 1970s, after Britain's accession to the European Community, rural policy meant agricultural policy. In the 1980s, however, and particularly towards the end of the decade, the progressive redirection of Community policy towards broader rural objectives has had an inevitable impact in this country (Hill, 1991). The publication in 1988 of the key report 'The Future of Rural Society' (CEC, 1988) signalled a major change into a broad-based rural policy approach in Europe.

In particular, the reorganization of the main Community Structural Funds, the Regional Development Fund, the Social Fund and the Guidance Section of the Common Agricultural Policy, in the late 1980s, resulted in the recognition of five sets of Objectives for the direction of Community programmes, three of which were defined spatially. Of these, Objective 5b regions (relating to areas deemed in need of rural development) were important for rural areas, although the original set of designations recognized very little of England and Wales. The rephrasing of the Objectives and the redesignation of relevant areas in 1993 resulted, however, in a significant increase in the area of rural England and effectively the whole of rural Wales, which is able to benefit from this initiative.

An increased European dimension to rural policy generally, and particularly to policy directed towards rural economic growth, would seem to be an inevitable emphasis towards the end of the century. Economic planning whether by local government or by specialist rural agencies is increasingly taking on a European perspective (Clout, 1993), while rural areas are likely to see an equivalent cosmopolitan flavour to local economic and employment structures.

8

COUNTRYSIDE RECREATION AND LANDSCAPE PROTECTION SINCE 1945

The post-war world turned out very differently in the assumptions and expectations of those who framed the country's policies for the years after 1945. Those relating to the rural environment and landscape were no exception.

During the wartime years the issues of scenic protection and the leisure use of the countryside were brought increasingly to the fore. One reason for this was that they had been emotive questions in the 1920s and 1930s, as described in chapter 4, and they had remained largely unresolved. Another reason was that as part of the propaganda drive to lift sagging spirits during the war, it became politically useful to restore morale by offering a vision of a renewed Britain in which the country's scenic and leisure resources were accessible to Everyman. Planning was in vogue and the resources of the countryside were no exception.

Hence by 1945 important points of principle were being established with regard to such matters as National Parks, nature conservation and related matters. Over the next few years these were progressively translated into statutory recognition by legislation and the setting up of appropriate public agencies to discharge new roles and functions. By the 1960s significant adaptations were being implemented, particularly in respect of recreation provision to cope with changing requirements. By the 1970s new landscape protection measures were being recorded as the parameters of an enlivened environmental debate made unexpected demands. By the mid-1990s the cumulative effects of 50 years of continuity and change had produced a situation much more complex than the policy makers of the 1940s ever expected.

In this chapter the focus is first on countryside recreation and the measures taken to realize the cherished dream of the National Parks movement. The progression covers the steps which led to the setting up of the National Parks Commission and the creation of ten National Parks in England and Wales between 1950 and 1957, its demise in favour of the Countryside Commission in 1968, and further developments over the last quarter of a century. The second section covers the plethora of measures followed in the interests of scenic protection, including such issues as the work of the Nature Conservancy, concern over the coast, land designations including Areas of Outstanding Natural Beauty and Environmentally Sensitive Areas, and the response to the unforeseen ravages to protected areas by forms of scientific agriculture. Chapter 6, which deals with post-war agriculture, also relates to some of these issues, so there are a number of points of cross-reference to pursue. The third section permits a synoptic impression of the whole post-war period, noting the shift of emphasis first from scenic protection to recreation, and then back to landscape concerns once more. The pattern established in other chapters is followed by adopting a chronological perspective to a review which now covers a period of half a century. In such a long embrace it is to be expected that long-term consistency soon breaks down, so the story is one of continual adjustment of policy to changing circumstances, marked by the interplay of political, professional and special interest group pressures.

COUNTRYSIDE RECREATION: NATIONAL PARKS, COUNTRY PARKS AND WIDER CONCERNS

National Parks

The origins of a policy towards National Parks have already been noted. There was first the identification of an issue – in this case an uncomfortable juxtaposition of two related questions, the protection of scenic quality and provision for access to the countryside for its enjoyment. Then followed an increasingly sophisticated articulation of pressure on politicians and policy makers in order for interest groups to secure their aims: civil disobedience by the masses and the emergence of a powerful lobby with respected leaders working to change attitudes. From the 1930s measures for policy formulation began to unfold – drawn up by local authorities with their private powers and by central government and its advisers.

In chapter 4 the story had reached the ineffective Access to Mountains Act, 1939. The advent of war brought an effective halt to speculation about countryside recreation, but from 1942 onwards the pace quickened (Cherry, 1975; Sheail, 1975; MacEwen, 1982; Blunden and Curry, 1990). At the risk of repetition, it is necessary to refer to the wartime happenings, as described in chapter 5, to appreciate the continuity of debate on National Parks and other issues from pre-war to post-war. There had already been a flurry of memoranda from amenity bodies on post-war planning, directed towards Sir John (later Lord) Reith in his capacity as Minister of Works and Buildings, an office which (to him at least) invited a responsibility for post-war reconstruction. The Standing Committee on National Parks recommended the setting up of a National Parks Commission, supplanting its earlier belief that National Parks could largely be secured by Planning Schemes, facilitated through the Town and Country Planning Act, 1932. Reith was receptive to this pressure, having already appointed (in October 1941) a Committee under the chairmanship of Lord Justice Scott to consider a wide range of rural matters, but, demoted in February 1942, he was replaced by Lord Portal as Minister of Works and Planning.

The Scott Committee *Report on Land Utilisation in Rural Areas* (1942) warmly embraced the idea of National Parks, believing their establishment 'long overdue'. The Committee recommended that the 'delimitation' of the Parks be undertaken nationally and urged the setting up of a national body to control them. Furthermore (hostage to fortune) the Committee thought that the main purpose of a Park was 'public recreation', and presumed that 'appropriate' camps would be provided in National Parks (paras 178, 180).

Lord Portal was subject to the same pressure as Sir John Reith, and when W.S. Morrison was appointed (the first) Minister of Town and Country Planning in December 1942, deputations were soon arranged. Debates about principles soon extended into arguments about administrative machinery, with government voices cautious or even hostile to the notion of a new authority being required for National Parks. In the meantime, however, pioneering work of almost a personal kind was afoot – John Dower, an architect/planner-civil servant, surveying possible National Parks areas during August and September 1942 for the Ministry of Works and Planning (Sheail, 1995). He was no stranger to the National Parks scene, having been a collaborator with the Friends of the Lake District in the 1930s (Sandbach, 1978). His national work continued, and over the next two years, although interrupted by bouts of ill-health, he laid the foundations for his comprehensive study *National Parks in England and Wales* (Dower, 1945). Its publication was against the background of ever-stronger indications that National Parks would figure in post-war reconstruction, the 1944 White Paper *The Control of Land Use* referring to the establishment of National Parks in national policies.

The Dower Report was the first comprehensive account of National Parks in England and Wales since the Addison Report of 1931. Dower considered that some 8,000 square miles in England and Wales constituted potential National Park areas. The following ten, totalling 3,600 square miles, were 'suggested National Parks', suitable and desirable for establishment in a first, five-year period:

The Lake District
Dartmoor
Pembroke Coast
Craven Pennines (Wharfe, Aire and Ribble)
Exmoor and North Devon Coast
Snowdonia
The Peak District and Dovedale
Cornish Coast (selected parts)
Black Mountains and Brecon Beacons
The Roman Wall

A further twelve 'reserves for possible future National Parks', covering 4,400 square miles, were for consideration at a later stage:

The Norfolk Broads
Dorset Coast and Heaths
N.E. Cheviots (Till and Coquet)
Swaledale Pennines (with part of Wensleydale)
Merioneth Coast and Mountains (including Berwyns)
The Elenith Mountains (Elan, Towy and Cothi)
North York Moors and Coast
Berkshire and Marlborough Downs
North Pennines (South Tyne, Wear and Tees)
Howgill Fells (upper Lune)
Plynlimmon
Radnor and Clun Forests

In addition a third list of 34 tracts specified other amenity areas. For the first time a 'shopping list' for England and Wales had been prepared from expert observation in the field.

Dower also established important principles in National Park provision. They should be national, rather than local, and (in the egalitarian climate of his day) for people 'of every class and kind . . . not for any privileged or otherwise restricted section of the population but for all who care to refresh their minds and spirits and to exercise their bodies in a peaceful setting of natural beauty' (para 13). Being national, their cost should fall on national funds and their administration should be the concern of an appropriate national body: if National Parks were to be provided *for* the nation, they should be provided *by* the nation.

Dower's work had been subject to a good deal of inter-departmental scrutiny during its preparation. There was a broad measure of support for it, though the Treasury expressed concern over costs. The Minister at the very end of the war, in May 1945, sought authority to set up a Preparatory National Parks Commission, but was pressed by fellow Ministers to reconsider. The equivalent Scottish proposals, stemming from a Committee chaired by Sir Douglas Ramsay which prepared findings parallel to those of Dower, did not require a Preparatory Commission and relied instead on the appointment of a Committee for further advice. Morrison fell in line, appointing Sir Arthur Hobhouse (a landowner and chairman of the County Councils' Association) to chair an appropriate committee. Dower's Report, shorn of its Departmental status, was treated as a personal Report to the Minister.

The setting up of the Hobhouse Committee coincided with a change of government, responsibility for National Parks now passing to Lewis Silkin, the new Labour Minister of Town and Country Planning. He supported their establishment, though there were weightier issues before him: the questions of compensation and betterment were still to be settled and his main Town and Country Planning Bill awaited introduction. In the event, although the political tide was still running strongly in favour of National Parks, the passage of time awaiting the Hobhouse Committee deliberations caused no great anxiety.

Hobhouse reported in 1947, recommending a total of twelve National Parks, to be created in three instalments over a period of three years:

First:	The Lake District, North Wales, the Peak District and Dartmoor
Second:	The Yorkshire Dales, the Pembroke Coast, Exmoor, and the South Downs
Third:	The Roman Wall, the North York Moors, Brecon Beacons and Black Mountains, and the Broads

Of Dower's list of ten, Hobhouse omitted one (the Cornish Coast) on grounds of likely administrative difficulties arising from its geographical shape, but added three. These included two from Dower's reserve list (the Broads and the North York Moors) and one from the list of other amenity areas (the South Downs). Hobhouse also provided a reserve selection of 52 areas of high landscape quality, scientific interest and recreational value, to be known as Conservation Areas.

On the question of National Park administration, Hobhouse proposed a Commission consisting of a chairman and eight members. For the Parks themselves Hobhouse advocated Park Committees, with a composition not normally exceeding 35 in number, the chairman to be appointed by the Commission and the membership split half and half between local authority representatives and those appointed by the Commission. The question as to how far National Parks were local or national would simmer for many, many years.

National Parks were once more in the government's lap, and Silkin, with no Cabinet seat and head of a young and immature Ministry, was subject to conflicting pressures. His Departmental Permanent Secretary feared the competition from an independent Commission in matters of country planning. The Treasury was anxious about the financial implications of a new Commission. The very objectives of the Commission came under scrutiny, and the conflict over central and local interests remained unresolved. It would have been easy to seek a compromise, perhaps in the form of a Commission for Rural England (a suggestion that Evelyn Sharp, the Deputy Secretary, once put to him) (Cherry, 1975). But Silkin came under pressure from the Standing Committee to stand firm and Cabinet colleagues, notably Hugh Dalton, then Chancellor of the Duchy of Lancaster, gave important support. Dalton came from the coterie of ramblers who had fervently sought access to private land; their world is well captured by Tom Stephenson's reminiscences in *Forbidden Land* (1989). The Bill was introduced in 1949; the National Parks and Access to the Countryside Act received the Royal Assent in December, the legislation *inter alia* making provision for the establishment of a National Parks Commission and for the designation of National Parks.

Lewis Silkin, introducing his Bill at the Second Reading, described it as a 'people's charter for the open air, for the hikers and the ramblers, for everyone who lives to get out into the open air and enjoy the countryside. Without it they are fettered, deprived of their powers of access and facilities needed to make holidays enjoyable. With it the countryside is theirs to preserve, to cherish, to enjoy and to make their own'. (*Hansard*, col 1493, 31 March 1949, quoted in Cherry, 1975). This was the language of the 1930s, but much was soon to change.

Silkin's constituency was abolished in a Parliamentary Boundary Commission Review, and following the 1950 General Election (which returned Labour to power though with a greatly reduced majority) he was raised to the peerage as Baron Silkin of Dulwich. Hugh Dalton became Minister, and with his appointment, town and country planning became for the first time the responsibility of a Cabinet Minister, though his title changed to Minister of Local Government and Planning. After the General Election of October 1951, which returned the Conservatives to power, Harold Macmillan became head of a newly renamed Ministry of Housing and Local Government.

Immediate political change meant no departure from the National Park programme, the Conservatives implementing the Labour government's legislation. Ten National Parks were designated between 1951 and 1957: the Peak District, the Lake District, Snowdonia and Dartmoor (1951), the Pembroke Coast and the North York Moors (1952), the Yorkshire Dales and Exmoor (1954), Northumberland (1956) and the Brecon Beacons (1957). From Hobhouse's priority list only the Broads and the South Downs had been excluded. A total of 5,258 square miles, equivalent to 9 per cent of the total land area of England and Wales were designated National Park land. A quarter of a century had elapsed since the Addison Report of 1931; years of pressure from interest groups, Ministerial investigation and reports from appointed bodies had come to fruition.

But subtle changes were afoot. The government's commitment to the National Park cause was increasingly questioned, local-central antagonism persisted, and new countryside issues came to the fore. By the end of the

Landscape in the National Parks. Langsdale in the Lake District National Park. A characteristic view of an upland landscape which has been protected and managed since the designation of National Parks. (By courtesy of the Countryside Commission)

1950s Silkin's parliamentary enthusiasm for his 'people's charter' seemed to belong to another age.

One disappointment was that National Parks for Scotland failed to ensue. Certainly there were some important differences between Scotland and England and Wales in terms of land law, resources, and the whole approach to National Park protection, but, with regard to landscape and scenic value, National Parks north of the Border seemed particularly appropriate. It was expected that a Scottish Bill would follow legislation for England and Wales, but with postponement the matter fell into abeyance and finally lapsed.

Difficult issues in Park administration served to sour relations with both local government and local amenity groups; whatever political good-will towards National Parks there had been, was scarcely maintained. The 1949 Act provided that where a National Park lay within the area of more than one local planning authority then a joint planning board should be set up, though the Minister could dispense with this if it was expedient to do so.

The Peak set up a Joint Board and appointed its own Planning Officer, and the Lake District followed suit with the administrative formula of a Joint Board, but decided to use the joint services of three county planning officers (for Lancashire, Cumberland and Westmorland) rather than appoint another officer. In Snowdonia the three County Councils pressed not to have a Joint Board; Dalton declined to accept this, but with a change of government Macmillan accepted the establishment of a Joint Advisory Committee. In Dartmoor there was dispute as to the composition of the Park Committee, fuelling old anxieties about the proportion of local and central representation. In Exmoor, the Commission's proposal for a National Park was fought by Devon and Somerset on the grounds that it was already adequately protected by the Town and Country Planning Act, 1947. As a concession Exmoor was allowed a Joint Advisory Committee rather than a Board, a procedure also followed in the Yorkshire Dales.

The National Parks had got off to an uncomfortable start, and the Commission was rebuffed when it pressed for greater financial support. But there were other anxieties for the vociferous National Parks lobby, largely over the government's record in protecting amenity. National Park designation seemed to promise the State the role of a national guardian of amenity, but all too often government appeared ready to draw back and appease a developer. There were some classic examples to record (Cherry, 1975). In 1958 the Minister of Power supported by the Minister for Housing and Local Government, approved an application by the Central Electricity Generating Board for a nuclear power station at Trawsfynydd in Snowdonia. Two hundred acres of uncompromising building and attendant overhead transmission wires were strongly opposed by the Commission, to no avail. Around the same time, again in the national interest, planning approvals were given for an oil discharge installation on the south shore of Milford Haven, just within the Pembroke National Park, and an oil refinery on the north shore, two thirds of which would be within the Park. In another 'sell out' to National Park principles (as the lobby saw it), a Ballistic Missile Early Warning Station was permitted at Fylingdales in the North York Moors. There were many other lesser examples, nation-wide, when National Park interests had seemingly been set aside to allow developments alien to the Park concept.

Country Parks

But in any case, other circumstances were changing too, and it was increasingly clear that National Parks would have to take their place in a rather different order of things. By the beginning of the 1960s demographic factors were heralding a major population increase and economic prosperity supported a dramatic rise in car numbers. By 1950 the number of cars on British roads had recovered to 1939 levels; by 1959 the number had doubled and by 1961 it had trebled. The leisure demands of the 1930s had been articulated by the hiker and the rambler; by the 1960s the mass recreationist was a motorist, who sought out not the open down and moorland, but the coast and the urban fringe countryside, typically within the 30-mile, one-hour drive. A new set of needs was becoming paramount. Quite apart from the countervailing attraction of overseas holidays, the use of the British countryside was being put to different purposes. The demand for access to open land now moved to car parks, lay-bys, beauty spots and 'honeypot' facilities.

It was time for a change in countryside recreation planning. The former flagship of National Parks now had competitors and a change of emphasis was required. The National Parks Commission had diminished effectiveness, its programme of Park designations coming to an end in 1957, and the impression could not be avoided that its major head of steam had gone. Government was not prepared to inject additional money into the Park system, local authorities themselves were poor spenders, administrative reform made no progress so long as the difficulties over local-central representation could not be resolved, and the Ministry and the Commission maintained an uneasy relationship. It became apparent that a new National Parks Act, and with it a new National Parks Commission was required, less amateur and more expert in composition.

A log jam of indecision was broken with a change of administration, Labour winning the General Election of 1964. Both the Labour and Conservative Party Election Manifestos emphasized the promotion of outdoor recreation, the Conservatives promising a Countryside Commission with resources to care for the countryside and coast. The new government took advantage of opportunities for change. A new Department, the Ministry of Land and Natural Resources (MLNR), assumed responsibility for National Parks. The Ministry of Housing and Local Government retained a planning function in the Parks (the Secretary of State for Wales in Wales and Monmouth). The new MLNR Minister, F.T. Willey, moved quickly and his Department's White Paper, *Leisure in the Countryside*, was published in February 1966. Both the new circumstances and the new intentions were made clear:

> Now that there is more money, more leisure, and above all there are more cars, if the original intentions of the National Parks and Countryside Act, 1949, are to be fulfilled, and if the National Parks are to retain their distinctive character, then it is essential to make new provision for the enjoyment of the countryside elsewhere both to meet public demand and to relieve pressure on remote or outstandingly beautiful places. (para 7)

It was apparent that a forthcoming Bill would essentially be a measure for maintaining legislation for National Parks, extending grant aid to facilities both within and outside National Parks, and for introducing a new facility in a form previously referred to as 'countryside recreation sites', but now known as 'country parks'. There was certainly a shift of emphasis from National Parks to a countryside-wide remit, but any hope that a comprehensive countryside policy was envisaged was effectively scuppered in the division of responsibility between (at least) two Ministries. This problem was at least overcome when in February 1967 MLNR was abolished and integrated with MHLG, and the new Minister, Anthony Greenwood, proceeded with the Bill. After an uncontentious passage the Countryside Act reached the Statute Book in July 1968; the National Parks Commission was abolished and replaced by the Countryside Commission.

The principal legislation for National Parks had survived nearly 20 years. The 1968 Act widened the scope of countryside recreation measures from the single focus of National Parks and the new Commission set about its task with commendable energy. There was an initial alacrity by a number of enthusiastic local authorities and landowners who took up the provision of site-based facilities, notably country parks and picnic sites, in a pattern around the major centres of population, as the 1968 Act envisaged. A flurry of designations soon tailed off, but over a 20-year period (i.e. by 1988) 220 country parks and 264 picnic sites had been provided. This at

The management of outdoor recreation. The Visitor Centre at Sherwood Forest Country Park, Nottinghamshire, one of the 'honey pot' parks set up after the 1968 Countryside Act. (By courtesy of the Countryside Commission)

least shifted the geographical emphasis in recreation provision from the National Parks to the urban fringes. Further amplification of this came with the Lea Valley Regional Park in East London, created by special legislation following pressure from the Civic Trust, and the Colne Valley Park in west London, the product of local authority collaboration.

Wider Concerns

But the concern of the amenity groups soon returned to National Park issues (if indeed it ever left them). In 1971 a government Committee was set up under a Minister, Lord Sandford, to consider whether National Parks had fulfilled their purpose, and to make recommendations for future policies. Sandford, reporting in 1974, reviewed some familiar issues (the conflict between recreation and conservation and between the Parks and other public agencies, for example) and highlighted one relatively new concern (moorland ploughing) to accompany afforestation as a long-standing enemy of scenic preservationists. But it was in the failure to plan effectively in the Parks that Sandford made his most telling points. Statutory development plans and structure plans lacked the follow-up management programme, and National Parks needed their own non-statutory plans.

To some extent Sandford was overtaken by the local government changes taking effect in 1974, as a consequence of the Local Government Act, 1972. The Boards for the Peak and Lake District Parks were reconstituted, while for the rest each Park would have its own committee, with an appointed National Park Officer, with the duty of preparing a National Park plan.

Much of the rest of the National Park story is intertwined with other issues, notably those concerned with environmental management and the Wildlife and Countryside Act, 1981, reviewed later in this chapter. The new Countryside Commission attempted to renew the National Parks programme, albeit with little success at first. In 1972 the Secretary of State for Wales declined to approve a Cambrian Mountains National Park, opposition from local farmers and residents proving too strong, and in 1976 the Commission's proposed East Anglian Broads National Park was withdrawn in the face of local opposition. However, in recent years the Norfolk Broads (1989) and the New Forest (1992) have been given National Park status, but in their cases the standard administrative machinery of a Park Board acting as a planning authority has been omitted. The boundaries of the New Forest Heritage Area (to give the new nomenclature) have been considerably expanded beyond those of the medieval forest, in order to protect the core against inappropriate development in the outlying areas.

The National Park history has been recounted in some detail, so far to the exclusion of other interesting developments in the post-war provision of countryside recreation facilities. But the emphasis highlights an important twentieth-century feature. For centuries the countryside has served as a setting for leisure. For the most part it was for privileged groups, but for the general public common land was important and footpaths and bridleways were time-enshrined public rights of way. By the nineteenth century however a good deal of the countryside had become privatized, with public access restricted by statutes, bylaws, enclosure, seizure and a variety of restrictions. In the twentieth century the pattern has been to break down the exclusive system, and post-war planning was expected to play its part in this: hence the allocation of land for particular purposes, the provision of leisure facilities and the means of access to them, and the management of the countryside resource (from sand dunes to moorland and from river valleys to footpaths) for community benefit.

National Parks became the early flagship of this planned enterprise, and, because of their emotive history over 60 years in England and Wales, they have retained the capacity to bring inflamed passions to bear. The use of the countryside for mass recreation poses particularly difficult problems. For one thing, the greater part of the countryside is in multiple use, including agriculture, water storage, mineral extraction, forestry and sport. For another, it is in multiple occupancy – landlords, tenants, private consortia, the military and public agencies. Furthermore, it is subject to multiple interests – not least those who would conserve and those who would develop. The practice of countryside recreation planning therefore became very political; it proved to be about the resolution of conflict, and those conflicts often proved at their most intractable in the National Parks.

Post-war recreation provision by the public sector in the countryside has not of course been confined to National Parks, nor indeed country parks. Additionally there has been water development by a variety of agencies, fostered briefly in the 1970s by the Water Space Amenity Commission, canal enhancement by the British Waterways Board, and the work of the Tourist Boards. Above all, however, there has been the contribution of the Forestry Commission in the provision of camping sites, car parks and picnic areas in the two Forest Parks of Dean and Snowdonia. England still has only 7.3 per cent of its land surface in woodland, the lowest proportion of any EU country except Ireland, with 5 per cent (Wales has 11.6 and Scotland 12.6 per cent). Concern about the greening of the urban environment led the Countryside Commission to propose a major initiative to improve countryside recreation opportunities for disadvantaged urban areas. Extensive urban forests were suggested for the outskirts of Walsall, Jarrow and Havering (East London), each forest site covering about 50 square miles. Later, in 1989, the Countryside Commission and the Forestry

Commission jointly amplified the proposals to create twelve community forests on the fringes of major conurbations, sites including Cleveland, Sheffield, West Manchester, East Liverpool, Neath/Port Talbot and North Bristol.

Complementary to these various measures for outdoor recreation there is a whole raft of private provision for golf courses, theme parks and facilities for countless activities from fishing to hang gliding and from field sports to country zoos. The range of provision is hugely varied as Patmore's (1983) informative review suggests and as has been reinforced in recent years by useful studies by Glyptis (1991, 1993*a*,*b*) and Curry (1994).

The protection of facilities for rural walking has been a major preoccupation throughout the post-war half-century. The question of footpaths had figured prominently in the Dower Report and at least two problems were well known: the legal provisions for establishing proof of right of way, and liability for upkeep of rights of way. In the early months of the Hobhouse Committee, its terms of reference were extended to include consideration of the general law concerning access and rights of way and in July 1946 a Special Committee On Footpaths and Access was

The post-war growth of recreation management. A view of the long-distance footpath created by the Countryside Commission along the River Thames near Radcot, Oxfordshire. (By courtesy of the Countryside Commission)

set up, also under the chairmanship of Sir Arthur Hobhouse. Its report recommended that a complete survey of all rights of way should be undertaken and completed within a period of four years by County and County Borough Councils. The 1949 Act incorporated this requirement and although the survey proved a protracted process, the country is now covered by 'definitive maps' which give conclusive evidence of rights of way. Battles have continued between the public, farmers and local councils with counter-allegations of neglect and damage to agricultural practices, but by and large the provision of a signposted and maintained network of public paths has been a major benefit of the post-war years.

The National Parks Commission and its successor have augmented this network through the creation of new, long-distance footpaths. They proved very slow to put in place, but some lengthy scenic routes now include the Pennine Way, Offa's Dyke Path, Pembroke Coast Path, South Downs Way, South West Peninsula Coast Path, Cleveland Way, North Downs Way, the Ridgeway (Berkshire and Wiltshire Downs), the Wolds Way and Peddars Way).

SCENIC AND ENVIRONMENTAL PROTECTION

It has already been observed that the twin features of post-war countryside planning – scenic protection and outdoor recreation – have been closely linked. But they have to be treated separately if much sense is to be made of a complex period of government action in this field. The matter is further complicated by the fact that the question of scenic protection proved particularly complex, as the following section indicates. Some simplification is attempted by introducing sub-headings.

Nature Conservation

The question of nature conservation was taken up by an off-shoot of the Hobhouse Committee (the Wildlife Conservation Special Committee) and chaired by one of its members, Sir Julian Huxley. Their Report, *Conservation of Nature in England and Wales*, was published in 1947. With the setting up of the Nature Conservancy (the national Biological Service proposed by the Special Committee) a fundamental separation from the National Parks movement occurred. At this juncture a regrettable split appeared between the 'hard science' element of environmentalism and the 'soft social science' of amenity interests and recreation.

This new State involvement had been some time in gestation. As Sheail (1976*a*) recounts in his history of nature conservation in Britain, a series of attempts had been made over many years to establish machinery for the preservation of amenity and wildlife. The 1940s provided the occasion when demands for the more rational use of the nation's resources met a positive response. As early as June 1941 a Conference on Nature Preservation in Post-war Reconstruction had been convened by the Society for the Preservation of Nature Reserves, attended by over 30 organizations, and a memorandum from the Conference attracted wide publicity for their cause. Subsequently the Scott Report (1942) gave support to the notion of national nature reserves, and with government encouragement a Nature Reserves Investigation Committee (NRIC) was established in 1942 as a follow up to the Conference a year earlier; their report was submitted in memorandum form in 1943.

Meanwhile the promotion of nature reserves was swept along in train with Dower's work on National Parks, and the British Ecological Society proceeded to draw up a list of vegetation types and areas which should be protected in the reserves. The NRIC proved indefatigable in promoting survey work and by 1945 a list of proposed nature reserves had been drafted. The setting up of the Wild Life Conservation Special Committee, contemporaneously with the Hobhouse Committee on National Parks, was made possible by the enthusiastic, collaborate work undertaken over the previous five years. Huxley was appointed chairman, though when he was appointed to UNESCO in May 1946, his vice-chairman, A. G. Tansley, succeeded him. A.E. Truman and J.A. Steers were appointed special advisers on geology and coastal physiography respectively.

The Committee drew up a list of 73 proposed national nature reserves for England and Wales, which included nearly all those previously listed by the NRIC. They were placed in no order of priority; each was claimed to be roughly equal in value and, it was argued, had to be preserved to secure a balanced representation of the major habitats of the country. In total they occupied an area of 70,000 acres and ranged from 20 to 3500 acres in size. The Committee also proposed a number of 'scientific areas', some of which overlapped with Dower's list of National Parks. They also listed a large number of geological 'monuments'. There still remained many hundreds of sites which could not be included on other lists because of their small size: they were termed Sites of Special Scientific Interest.

National Parks and nature conservation went their separate ways. The former remained the responsibility of the Minister of Town and Country Planning, the latter the province of the Lord President of the Council, who represented scientific concerns in the Cabinet. Nature conservation was to be the preserve of scientists, who quite early on demanded special privileges. With abounding confidence and total certainty in the merits of scientific method, the Council of the British Ecological Society had recommended in 1943 the establishment of an Ecological Research Council to be responsible for nature reserves, to undertake surveys of flora and fauna and establish an institute of terrestrial ecology.

In the event, the Huxley Committee proposed a Biological Service responsible for selecting, acquiring and managing nature reserves and for undertaking the necessary survey and research work; this Service would be attached to the Government through a Nature Conservancy Board. These recommendations were largely accepted and in 1948, with A.G. Tansley as the first chairman, the Nature Conservancy was set up. A royal charter set out the responsibilities:

> to provide scientific advice on the conservation and control of the natural flora and fauna of Great Britain; to establish, maintain and manage nature reserves in Great Britain, including the maintenance of physical features of scientific interest; and to organise and develop the research and scientific services related thereto.

Organizational changes subsequently took place in a rather tortured sequence of events. The first was in 1965 when the Nature Conservancy became a component part of a new research council, the Natural Environment Research Council. The second occurred in 1973 when the Conservancy was abolished and a new Nature Conservancy Council was created as a grant-aided body, responsible to the Department of the Environment; the research staff of the former Conservancy remained in the Research Council and became the Institute of Terrestrial Ecology. The third was in 1991 when, under the Environmental Protection Act 1990, the Nature Conservancy Council was split into separate bodies for England (English Nature), Scotland (the Nature Conservancy Council for Scotland) and Wales (the Countryside Council for Wales). This latter body also took over the Countryside Commission's responsibilities in Wales, leaving the Commission responsible for countryside policy in England only.

For the first quarter of a century relatively untroubled progress was recorded. A.G. Tansley thought that the number of National Nature Reserves in England and Wales was unlikely to exceed one hundred (Sheail, 1976*a*). By January 1975 a total of 99 had been established (rising to 174 in 1991), a figure complemented by other Local Nature Reserves and Forest Nature Reserves. The earlier concept of 'scientific areas' was not proceeded with, but Sites of Special Scientific Interest were, and by 1975 their number in England and Wales had risen to nearly 2500 (4352 in 1991). The voluntary bodies had been given a powerful umbrella under which to work, and the County Naturalists' Trusts flourished. Local authorities too were given a role, nature conservation finding a place in the planning system through development plan provisions.

Environmental Protection

But by the 1970s new issues were apparent and an increasingly vigorous environmental debate was being engaged. Nan Fairbrother's *New Lives,*

New Landscapes (1970) was an early warning of the societal and landscape changes that were unfolding. Westmacott and Worthington (1974) in a Countryside Commission paper, *New Agricultural Landscapes*, described the reduction of trees and general vegetative cover in the English lowlands. A further corroborative report followed from the Nature Conservancy Council (1977) in *Agriculture and Nature Conservation* which showed the sharp reduction in numbers and varieties of plants and animals.

Evidence regarding losses of natural heritage mounted as the years went by. Lowland herb-rich hay meadows were damaged by agricultural intensification. Lowland grasslands and sheep walks in chalk and limestone areas were lost by conversion to arable or to improved grassland. Lowland heaths on acidic soils were also being converted to arable or improved grassland, or afforestation. Ancient woodland was lost to conifer plantation or was being grubbed out to provide more farmland. Lowland fens were drained and reclaimed for agriculture. As described in chapter 6, new farming practices embraced the widespread application of fertilizers and herbicides; and farm rationalization led to hedgerow loss. Upland grassland, moors and heaths were taken for coniferous afforestation, hill land improvement and reclamation. Other habitat loss and drainage extended to inter-tidal flats and saltings, where reclamation for both agricultural and industrial purposes was taking place. Sand dune areas faced recreational visitor pressure. Lakes and rivers generally suffered from pollution, water abstraction, land drainage schemes and acidification. Another feature was the damage to limestone pavements in northern England, largely through the removal of weathered surfaces for sale as rockery stone.

The fate of the English hedgerow might be considered in greater detail. Found rarely in other parts of the world (they are present in Brittany, Normandy, Ireland, New England and Tasmania) many date from the enclosures of the old open fields during the eighteenth century, but there are many others much older. They lie at the heart of a conflict of interest represented by the farmer, the conservationist and the informal recreationist whose delight is scenic tradition. No less than 500,000 miles were thought to exist at the close of the Second World War, but today barely half remain. Government subsidies were given to farmers to grub up unwanted hedgerows in the name of agricultural improvement (tractors can be worked much more efficiently if they do not have a multiplicity of awkward corners to negotiate). The value of hedges as stock-proof barriers declined, and for the cereal farmer hedges harboured pests. The subsidy arrangements came to an end in the 1970s; now the incentive is to plant new hedges and care for existing ones. However, the total hedgerow length has continued to fall: from 341,000 miles in 1984 to 266,000 in 1990. Currently England and Wales are losing about 16,000 miles of hedgerow a year. In the

Conservative Party Election Manifesto of 1992 there was a promise to underpin the incentives with legislation. Government did indeed support a private members bill sponsored by Peter Ainsworth, requiring farmers to seek local authority approval before removing a hedgerow, but the bill was 'talked out' by MPs representing farming and landowning interests on the grounds that it took away farmers' rights to use their land in the most profitable way. Employment (farming), ecology and recreation: the simple hedgerow provides a classic example of a fearsome clash of interests.

Landscape changes on the scales described had inevitable consequences for wildlife, flora and fauna. The number of butterfly species contracted, reptiles and amphibians were endangered, breeding birds declined, the otter became rare, bats were at risk of extinction, and species of native flowering plants and ferns fell back sharply in affected areas.

Furthermore, large-scale urban-oriented ventures had their own implications for environmental conservation: the tide of bricks and mortar in an unprecedented building boom, road construction programmes, and industrial projects. The latter provided (as with National Parks) an example of how other considerations could override conservation principles. The Tees Valley and Cleveland Water Board gained approval in 1967 for a reservoir at Cow Green in order to provide water for Teesside; part of the site was in the Upper Teesdale National Nature Reserve and the remainder had been notified as a SSSI. It was in cases like these that the clash of value systems in the politics of planning was firmly recognized. Gregory (1971) instanced five examples of recent national importance where industrial development had impacted on amenity and the national environment: the Cow Green Reservoir (above), ironstone mining in Oxfordshire, power stations on the Trent, the North Sea gas terminal at Bacton and the Abingdon gas holder. The resolution of conflicting interests in amenity questions would haunt the rest of the century: planning decisions on amenity issues are matters of choice and priorities, where facts are not neutral, and decisions are arrived at by the inter-play of political power.

These new environmental and amenity issues, and the political contentions aroused, led to fresh legislative and organizational initiatives to cope with a situation which could not have been envisaged in the late 1940s. But first, mention must be made of the remaining complement of land use protection measures for two important tracts of the countryside: the coast and Areas of Outstanding Natural Beauty.

The Coast

As previously observed, the coastal areas had already been subject to damaging pressures from development before the war (Sheail, 1976*b*). In the enthusiasm for post-war planning, engendered from late 1941 onwards, it was obvious that the coast would receive attention. Many beaches had been despoiled during the war and a speedy return to coastal holiday-making was anticipated. The Scott Report (1942) in its scatter of recommendations for rural land use planning thought that 'the coast of England and Wales should be considered as a whole with a view to the prevention of further spoilation' (p. 59). In September 1942 regional planning officers were asked to undertake coastal surveys as part of their work. By 1943 the Ministry's interests had focused on scenic assessments, rather than the provision and siting of holiday camps, and the geographer J.A. Steers was recruited as adviser. His book *The Coastline of England and Wales* (1946) would mark him out as an authority. Between 1943 and 1945 he visited and reported on each stretch of coastline in England and Wales, continuing his surveys in Scotland between 1946 and 1953. Steers's reports helped to identify those stretches of the coast which should be protected and those unexceptional stretches which could admit development.

Although Steers also advocated seeing the coast as a whole, and not in any piecemeal manner, no central coastal planning authority was set up. The National Parks Commission held some coordinating hand, but in the main responsibility for coastal protection fell to local authorities. They were hard pressed to cope with the growing demands for holidays, homes and caravans by the seaside, as reflected in the National Trust's 'Enterprise Neptune' scheme in 1965 to raise £2 million for acquiring tracts of coastline. Within eight years the length of coast owned by the Trust almost doubled, from 185 miles to 360 miles. The Trust has since become the largest owner of coastline in England, Wales and Northern Ireland, now protecting 530 miles.

A report by the Countryside Commission (1968), based on work undertaken by the National Parks Commission between 1966 and 1967, suggested that, of a coastline length for England and Wales of 2,742 miles, nearly two-thirds (63 per cent) was covered by specific planning policies restricting new development, the greatest protection afforded in Pembrokeshire, Devon, Anglesey, Cornwall and Cardiganshire. One quarter could be regarded as developed or committed for development. Fifteen per cent of the coastal frontage was in one or more forms of protective ownership; nearly 5 per cent was held by government departments for defence or other purposes. These data encouraged the Countryside Commission in 1970 to propose the designation of 'Heritage Coasts', for the management of which special committees should be established. Government rejected both formal designation and the special administration machinery, but the term is applied to coastline of high scenic quality in structure plans and local plans. One-third of the open coast of England and Wales is now covered by 44 stretches of Heritage Coast.

But concern over coastal protection has continued. In 1992 the House of

Commons Environment Select Committee agreed the report of its inquiry into coastal zone protection and planning. The Committee found that current legislation did not provide an integrated or efficient framework for coastal protection and planning; consolidation and updating was recommended. The Committee also suggested a review of the roles played by the many organizations with an interest in the coastal zone in order to reduce duplication of responsibilities and improve coordination. Meanwhile policy guidance for coastal planning is provided by Planning Policy Guidance Note 20 (1992).

Areas of Outstanding National Beauty

Another important land area of scenic quality constituted those tracts which Dower had referred to as 'reserves' and 'other amenity areas' (Smart and Anderson, 1990). The former were substantial, but the latter either too small or isolated for national action. Hobhouse had the same notion of a supplementary list and first thought of them as 'Conservation Areas', in parallel with a designation adopted by Huxley's Special Committee on Wild Life Conservation. To begin with, Hobhouse thought of them as primarily for recreation, but this view was substantially modified to a more preservationist role. A list of 52 appeared in the Hobhouse Committee Report (1947).

The term had changed by the time the National Parks and Access to Countryside Bill was introduced to 'Areas of Outstanding Natural Beauty'. Their agricultural role was also emphasized. As with National Parks, their natural beauty was to be preserved and enhanced, but enjoyment by the public was not to be promoted. Hobhouse also thought that, while making the areas the responsibility of local authorities, Advisory Committees should be set up to assist in their management; however, the Act made no such provision. Procedures were simply that the designation of an Area should be agreed jointly by the National Parks Commission and the local authority and submitted to the Minister for final approval.

Hobhouse's Conservation Areas had been changed in name and function and downgraded in concept. In the event, perhaps rather more has been made of them than might have been expected. It was not until 1956 that the first AONB (the Gower Peninsula, South Wales) was designated, but a total of 25 had been confirmed by the end of 1968 when the Countryside Commission replaced the National Parks Commission. By the early 1990s 38 AONBs covered 12.8 per cent of England and Wales. They have been largely successful in resisting forms of building development, but they have not escaped landscape change emanating from scientific farming methods.

Landscape and Habitat

The environmental debate had quickened. A flurry of pronouncements at the close of the 1960s and in the early 1970s brought out the new concerns. Robert Arvill's *Man and Environment* (1967) offered a new planning text which looked at human impact on land, air, water and wildlife – an early testimony of the revised, post-war agenda. Sir Frank Fraser Darling gave the BBC's Reith Lectures in 1969, to which he gave the title 'Wilderness and Plenty'; they dealt with the relationship between people and their natural environment. Max Nicholson's *The Environmental Revolution* (1970) expressed an international concern about the vulnerability of man's natural environment. *Only One Earth* by Barbara Ward and René Dubos (1972), a report commissioned for the United Nations Conference on Human Environment, held in Stockholm, stressed the need for a global approach to pressing environmental problems. More or less at the same time three other 'landmark' publications expounded the principal ideas of the new environmentalism. *Limits to Growth* (Meadows *et al.*, 1972), a computer modelling of the consequences of current physical, economic and social trends in the world, forecast disastrous results for the earth's carrying capacity; it implied the need for zero population growth, as well as zero economic growth. 'Blueprint for Survival', a piece published in *The Ecologist* (Goldsmith *et al.*, 1972) advocated minimal ecological disruption and maximum conservation of energy and materials for future British society. Schumacher's *Small is Beautiful* (1973) dealt with the values inherent in currently-held philosophies which applied to man's relationship with nature.

These concerns provided a new intellectual context for those engaged in the issues of post-war land management in Britain. They were philosophical, in part speculative and mostly long-term. But it was the immediate impact of new agricultural practices which prepared a practical agenda for Britain by relating the question of landscape protection to the general concerns of conservation. The immediate issues included increased mechanization, rationalization of farm units giving larger fields and a loss of hedgerows, drainage of the wetlands, ploughing of formerly marginal lands, and greater application of herbicides and fertilizers. A cross-reference to chapter 6 at this point provides the context. The irony was that the conventional wisdom established by the Scott Report (1942) was that the essence of the British countryside was that it was farmed, and the best way to maintain its well-loved features was to keep it farmed in a traditional way; unconsciously the farmers and the foresters were the nation's landscape gardeners, and the agriculturalist was the guardian of landscape preservation. The belief was that there was no conflict between a healthy and efficient agriculture and the conservation of landscape and wildlife.

But the alien scale of agribusinesses and new farm technology proved capable of transforming the traditional landscape and disrupting fragile habitats. In due time the unwelcome change to the English countryside sharpened the anxieties of the amenity lobbies. One critical observer, Marion Shoard (1980) wrote of 'the theft of the countryside', alleging that it was under sentence of death, and the executioner was the figure traditionally viewed as the custodian of the rural scene – the farmer.

A further irony was that the changes were being experienced in some of the very areas designated for special protection – moorland, wetland and ancient grassland. One good example of an area to undergo landscape change as a result of different farming practices is the 250 square miles constituting the Somerset Levels, that low-lying tract between the Mendips and the Quantocks where eight sluggish rivers debouch into the Bristol Channel. In the past, summer grazing of cattle gave way to shallow flooding during winter and spring, to form ideal conditions for wildfowl and wading birds. New technology in the form of high-powered pumps permitted the speedy draining of the winter floods. This meant that sheep could be grazed in winter and cattle put out to pasture in the spring. The water meadows were drained and ploughed for crops or sown with mixtures of rye grasses, that fatten animals faster but crowd out other plants. Bird life fell back and wetland flowers became rarer. The Levels are but one example of a landscape transformed by new farming methods, but in many parts of England technologically efficient agriculture, financially cushioned by cereal price support mechanisms, soon showed itself capable of fashioning greater change on the landscape than since the destruction of the open fields in the enclosure movement.

The hostility to ploughing on National Park moorlands, first in Exmoor, brought matters to a head. Moorland ploughing on Exmoor was first apparent in the 1960s and the Exmoor Society, a branch of the CPRE, made unsuccessful efforts to amend the Countryside Act, 1968, to give the National Park Authority powers to control moorland conservation in designated areas. There was also concern about damage to SSSIs caused by changes in agricultural practice, hence attempts were made to oblige farmers to notify the Nature Conservancy of their intention to undertake operations which might be detrimental to scientific interests. However, an uneasy calm prevailed, close links between the NFU and the CLA ensuring that the farmers' viewpoints were well represented. A change came about in 1977 when a special report by Lord Porchester examined the loss of moorland to agricultural improvement. The moorland was the very landscape feature that had prompted the inclusion of Exmoor as a National Park in the first place, but the new farming practice could not be stopped as it was exempt from any form of public control. Porchester proposed Moorland Conservation Orders in order to enable the National Park authority to oblige farmers not to improve moorland, where a voluntary agreement could not be reached; compensation would be payable to the farmer.

In December 1978 the Callaghan Administration introduced a Countryside Bill, embodying within it many of the recommendations of the Sandford and Porchester Reports. This gave National Park Authorities power to make Conservation Orders; once-for-all payments equal to the difference between unclaimed and reclaimed moor, would prevent moorland loss. However, the Bill fell with the General Election in May 1979. In a change of government, Michael Heseltine's re-introduced Bill (November 1980), the Wildlife and Countryside Bill, made no such provision for compulsory back-up power on the grounds that Exmoor was exceptional and that no other National Park was similarly threatened. Research was to show that this was not so, and that North York Moors, Brecon Beacons and Northumberland were also affected (Parry, Bruce and Harkness, 1982). The Government resisted compulsory reserve powers, but as a compromise accepted the need for preparing and regularly updating moorland maps so that changes could be monitored.

Apart from this aspect, the Wildlife and Countryside Bill covered much of the ground of the Labour Government's earlier Countryside Bill. However, in addition it was decided to amend and consolidate existing legislation for the conservation of plants and animals. Moving the Bill as a non-controversial piece of legislation (as it was believed), it was decided to commence the Bill's passage in the House of Lords – a procedure reserved for Bills thought to require little discussion and amendment. In the event the Bill attracted a record of 1,120 amendments, trimmed to 560 for the Lords Committee stage, taking 13 sitting days. From Introduction to Commons Third Reading the Bill took longer to negotiate Parliament than any other Bill that session (Cox and Lowe, 1983). The major part of the controversy centred around which animals and plants were to be in the Schedules listing those which were to be protected, but it all indicated the heightened concern attending environmental and wildlife issues.

The pivotal principle of the 1981 Act (it received the Royal Assent in October) was the control of the conflict between agriculture, and wildlife and landscape through the mechanism of management agreements, which were to be voluntary and fully compensated. The Nature Conservancy Council and/or the National Park in question would notify landowners of the conservation interest of their land when particular activities were likely to damage that interest. If a compromise could not be reached, the conservation authorities were obliged to compensate the landowner for any profit foregone as a result of his activity being denied, a procedural shift in landscape management, from custodialism to compensation, of profound significance.

A very confused period began, complicated by other agricultural issues. The background was the growing realization that Britain was now self-sufficient in most temperate foodstuffs and that continued farm subsidies were generating huge, unwanted surpluses. There was resort therefore to a variety of mechanisms to extensify production, including to take land out of production ('set-aside' land), issues which are explored more fully in chapter 6.

The principle of landscape protection through designation was maintained through the establishment of Environmentally Sensitive Areas. Originating as a British policy initiative, they were introduced by the European Community in 1985 and implemented in Britain by the Agriculture Act, 1986. Six were designated in that year after an experimental Broads Grazing Marsh Scheme where Norfolk farmers within a small trial area were paid £50 an acre in exchange for not draining, ploughing or over-fertilizing their land, or overstocking. The six covered a range of environmental conditions: grazing marsh (Somerset Levels and the Broads), grassland (Pennine Dales and South Downs), small scale field systems (West Penwith) and afforestation (Cambrian Mountains). Further ESAs have been designated in a substantial expansion of the programme. There are now 22 in England, covering 2.5 million acres where farmers are given financial inducements to manage and enhance traditional countryside features including heather moorland, chalk grassland or water meadows. The sums paid range from £10 to £300 per acre, depending on the profit foregone by refraining from intensive farming methods. The ESA programme also extends to Wales, Scotland and Northern Ireland.

The extent of the programme is perhaps surprising, as an interim evaluation of its success could only be cautious (Gaskell and Tanner, 1991). ESAs represent the addition of yet another designation to what is already a highly complex system of policies which set out to achieve social objectives and prevent environmental deterioration in areas of high amenity value. Agreements to be made with each farmer, and the arrangements for monitoring, are likely to imply high administrative costs. As ever with designations, boundary delineations can be problematic, and the voluntary approach to take up will limit environmental effectiveness. Overall, ESAs can only be a partial and temporary solution to the problems of landscape change, and at high cost, but their effectiveness in maintaining specific landscape features for a limited time cannot be denied.

RETROSPECT

In matters of the countryside, notions of the post-war world were shaped crucially by the experiences and unmet aspirations of the pre-war years.

This chapter has discussed prevailing sentiments regarding a raft of issues concerning landscape preservation (including wildlife, habitat and ecology) and greater access to and use of, the countryside for recreation purposes. The various issues were closely connected; amenity and recreation groups had shared sympathies, though their ultimate aims could be seriously in conflict.

The preservation ethic was born of a countryside in change: a workforce leaving the land, communities in decline, a fragile economic base, and a humanized, lived-in landscape facing an uncertain future before a brash assault from scarcely regulated development pressures. Yet a basic scenic quality remained, resting on a wonderfully varied geology, topography and flora and fauna, a heritage to be defended for future generations at all costs. The Scott Committee Report (1942) was written at a time when it was commonplace to write of a deep love for a countryside which was

> like a multicoloured chequer board. Its chief characteristic is its attractive patchwork appearance, with an infinite variety of small, odd-shaped fields of brown ploughland or green pasture bounded by twisting hedges, narrow winding lanes, small woodlands and copses and isolated trees and hedgerow timber; quiet streams and placid rivers: the whole giving place in remoter parts to open moorland or rolling downs. (para 11)

Nearly 40 years on the same evocative descriptions were being offered: 'a countryside of matchless charm, peopled by badgers, skylarks and nightingales, scattered with bluebells, poppies and cornflowers and studded with oak, elm and hawthorn' (Shoard, 1980). For the whole of the post-war period this preservation ethic has endured, partly sustained by romantic, pastoral myths, but in the main resting securely on a sustained appreciation of scenic quality and an historical, farmed landscape of intimate charm. Without these features, it has been regularly asserted, the country is the poorer; with them there is cultural strength, depth and tradition. This persistence of a nostalgia for rural life and landscape, which has raised the countryside to an idealized status, has recently been examined by Bunce (1994).

Against the background of inter-war concerns (the shapers of post-war policy were essentially the agitators and protagonists of the 1930s), the post-war outlook was conditioned from four particularly influential quarters: Scott, Dower, Hobhouse and Huxley. The Scott Committee (1942) argued that the continuance and revival of the traditional mixed character of British farming ('a healthy and well balanced agriculture') would have two important consequences. First, it would help to preserve the countryside because farming would retain those features which gave it charm and character; and second, it would help to resuscitate village life through an improvement of housing and general living conditions and an equalization of economic, social and educational opportunities. The

Dower Report (1945) and the Hobhouse Committee Report (1947) sketched the basis for National Park provision, listing areas for protection and recommending arrangements for their organization and administration. The Huxley Committee (1947) made recommendations on the conservation of nature.

Out of these deliberations two national agencies came into being: the National Parks Commission and the Nature Conservancy, and they and their successors have been permanent features of the post-war scene. The scientific service has always been the weightier of the two; scientific 'proven' criteria were always plausible when applied by experts to matters of nature conservation, whereas the amenity lobby was obliged to rely heavily on judgement and emotion and was more fickle and fluid in changing circumstances.

The work of Scott, Dower, Hobhouse and Huxley provided the early intellectual infrastructure for landscape and environmental policy for the post-war countryside. However, from the late 1960s the input of ideas came increasingly from much wider concerns, stemming from anxieties about an increasingly exploitative and wasteful period of human history, focusing on people and their relationship to the natural world. By the 1970s the issues which had formed the planning agenda thirty years earlier had moved on, and long-established policies, and methods for implementing them, were increasingly discredited. As Graham Moss (1981) later put it:

> As an administrative system, planning in Britain has been outstanding, but as a public ethic of responsibility, it has clearly failed. All too often . . . planning has been part of our modern throw-away society, treating land as just another used commodity to be discarded. (p. 207)

The vigour of the environmental movement, enlivened by radical green politics, dependent not so much on expert opinion (as in the past) but on the lobbying power and commitment of special interest groups, came to dominate the countryside scene in the last quarter of the century. Beset by the anxieties and issues which constituted the 'roots of modern environmentalism' (Pepper, 1984), the context for countryside planning significantly changed in this period. It began perhaps with a burgeoning population growth in the 1960s which fuelled speculation about future pressure on land in a crowded island. Meanwhile the results of economic growth could be seen in the (needless?) exploitation of natural resources; and the demands of modern technology (not only nuclear and other power generation, but also oil and gas processing) were held to be excessive. Stewardship and the husbanding of natural resources for further generations ('sustainability' became a key word in the 1990s) was regarded as imperative. Greater attention to the protection of flora and fauna and wildlife generally was urged in a recognition of the importance of what survives from one generation to another. In an extension of this concern, archaeological protection has been sought, Fowler (1987) arguing that the cultural heritage of sites and monuments is analogous to the natural heritage.

Over the last two decades, then, a very different context for action has emerged, with targets and policies embracing those of the 1940s, but going much beyond them. Meanwhile the planning system put in place after the war remains in its essentials. At macro level the post-war system for landscape protection and conservation relied heavily on the principle of designation – a site- or area-specific process of land allocation, reservation or protection for a particular purpose. At micro level the system then depended on appropriate management schemes and processes relative to particular areas or habitats (as described by Green (1985) for example), but that detail does not concern us here. The designatory system had its merits; it has been well understood by a technical and lay public, it has proved consistent and relatively even handed, and has worked well with the land use planning system, as developed statutorily after 1947. National Parks, AONBs, ESAs, Nature Reserves and SSSIs have been easily assimilated into the provisions of development plans and structure plans, while designations of purely county significance have amplified the extensive range of land protection zones.

But over time the designation system by itself failed to provide enduring safeguards. Particularly where the reason for protection was purely an amenity purpose, other overriding needs for development could all too often be found, and even where nature protection was at stake, preservation principles could not in themselves be sacrosanct. Even the heavily protected National Parks show considerable change in their traditional features. Over the last two decades agriculture and forestry have caused loss of heath and moorland, rough pasture, hedges and dry-stone walls. An area of heath and moor equivalent to about one thousand football pitches has disappeared each year on average, while conifer forest has expanded at about the same rate (Countryside Commission, 1991). Over a 15-year period from the mid-1970s to the late 1980s improved pasture increased by 26 square miles, and conifer forest by 51 square miles, grass moor by 19 square miles, lowland heath by 12 square miles and bracken by 7 square miles.

Meanwhile, day-to-day development control measures by local authorities have failed to prevent a rash of unwelcome features in the landscape, considered highly intrusive to sensitive environments. Development by government departments, military establishments and public agencies has been widely criticized, the electricity supply industry perhaps a major culprit, as Sheail (1991) describes in his history of the CEGB. Road improvement schemes which impact on high quality landscapes have attracted hostility, as seen in the examples of the A66 in the Lake District, the Okehampton Bypass (Dartmoor) and the cutting of Twyford Down, east of Winchester.

Designation is of course only an act, a declaration of intent, at a particular point in time. As a device for protection, designation is only as strong as the merits that lie behind the measure when applied. But circumstances change: National Park designation for recreation purposes became less important when the geographical spread and variety of provision made policies covering other areas of the countryside seem more important. On the other hand nature conservation, at risk from unregulated agricultural practices, did require additional, area-specific designation, as in ESAs.

One of the dominant impressions, however, of nearly 50 years of post-war planning for landscape and nature protection, amenity enhancement and recreation provision is how integrally ill-adjusted the landscape protection system actually is. From at least a semblance of an interrelated, comprehensive approach in the late 1940s, there has been little progress to record. Town and country planning legislation to all intents and purposes failed to provide control over the design and siting of farm buildings or agricultural practices, an omission which failed to withstand huge, unregulated landscape change and the intrusion of alien design and unsympathetic materials in building construction. The Forestry Commission, reconstituted in 1945, was slow to diversify its role from forestry enterprise into amenity, conservation and access. The scientific service for nature protection was the responsibility for one arm of government; the amenity and recreation interests were another's concern. Nature conservation and National Parks followed divergent tracks. The National Parks Commission ran out of steam, and the Countryside Commission came near to being disbanded in 1980. And where stands rural tourism today?; outdoor recreation is currently more about economic growth than countryside protection. Sectional interests seem paramount; Blacksell and Gilg (1981) in their study of rural issues in Devon were led to conclude that 'The Countryside Commission is too small and politically weak, MAFF too often identified with the farming lobby, and the DOE, too diverse and urban oriented' (p. 223).

Perhaps the surprise is that the protective system has held up as well as it has. The reason for this may well be the resilience of the planning system itself. Its essential characteristics (a plan-making capability, a responsibility for control of development by democratically elected councils, and the exercise of local discretion in arriving at decisions), all made for adaptability and flexibility. It was as well because one of the features of post-war planning has been its role of conflict mediation. Many issues in landscape, conservation and amenity planning are all about the conflicts within and between value systems – not just in England and Wales but throughout the developed world, if we follow Robinson (1990). The abiding questions concern protection of land for whose interest, whose interest should prevail and who decides that they should prevail? The growth in importance of amenity and preservation groups against the power of the expert, bureaucracies, landed and commercial interests and government has provided a new element in the system of checks and balances that characterises our present democracy, as we are reminded by Lowe and Goyder (1983). The planning system is just one, but an important absorbent and filter in mediating the conflicts that arise.

And so an imperfect system has endured, and the result is not all that unappealing. The countryside for the most part is remarkably well protected, with almost a surfeit of statutory and non-statutory protective designations, though without the arrangements for protection becoming a barrier to change. Moreover, the countryside is well used, a variety of interests co-existing without great friction. Indeed some features of rural life positively support others; areas rich in wildlife for example attract tourism, helpful to remote districts with fragile local economies. As far as outdoor recreation is concerned, the leisure boom has been sustained.

The pressures of recreation. Tarns Hows in the Lake District National Park. Walking the hills, which grew especially in the inter-war years (see page 50) has continued its popularity to the end of the century. (By courtesy of the Countryside Commission)

Participation rates in virtually every outdoor activity have increased and, by and large, providers have coped well in meeting demand; pressures have been absorbed remarkably satisfactorily.

No doubt it can be assumed that even the latest contentious problem (the impact of golf courses on landscape) will be taken in its stride; 50 years of landscape and amenity planning have been rather like that. Claims and counterclaims about the merits of the new boom in golf have typified the conflict of interests. It is alleged that an overproduction of courses despoils the landscape through the creation of an 'alien' terrain, extends the use of pesticides for preparation and maintenance of the greens, uses valuable water during droughts causing inconvenience to local people, and proliferates unwelcome forms of development in the form of new buildings, country clubs, access roads and the ubiquitous motor car. With more than 270 courses built in Britain since 1985 and with planning applications probably four times that number being considered during 1993, the issue is a lively one (though the bubble has undeniably burst now). But golfing authorities deny that courses are environmentally hostile; intensive agriculture has been far more damaging. The amount of fertilizer used is less than on agricultural land (about half the area of a golf course is rough and uncultivated). With good landscape design a course need not give an over-manicured, urban impression, more trees can be planted, wildlife encouraged and fauna enriched. The emotionally charged debate is typical of the landscape/recreation/development agenda over the last 70 years.

9

RURAL COMMUNITIES SINCE 1945

In 1943 the eminent agricultural economist C.S. Orwin conducted a comprehensive survey of an area of north Oxfordshire with a team of researchers from the Oxford Agricultural Economics Research Institute. The study was subsequently published in 1944 as *Country Planning: A Study of Rural Problems* (Oxford AERI, 1944). Though long out of print, it remains the classic study unhindered by excessive romanticism, of English rural communities in the middle of the twentieth century. Together with a well-produced film made for the Central Office of Information in 1946, these two records provide an obvious starting point for this chapter.

Though produced in wartime, Orwin's study barely touches on the immediacy of war. Instead he sets the rural community firmly in its historical context. The rural areas appear as *residual* areas, suffering from years of neglect and economic depression in the face of urbanization and industrialization which had sucked the country dry. But despite this clear historical scene-setting, Orwin's study could in no way be accused of showing the characteristics of much later rural writing and comment – the backward-looking sentiment which Howard Newby so aptly called 'rural retrospective regret' (Newby, 1979). Orwin set a practical agenda and, in analysing the state of economy and society, gave a wide ranging prescription for the future of the countryside.

At the start of his book Orwin lays out what he sees as four key problems of the countryside. These bear repeating for they provide a perceptive mid-century context for this chapter (and for others in this book):

> These, then, seem to be the problems of country planning. First, how can rural industrial life be organised – and farming in particular – so that it may give better returns both in goods and services, while providing more opportunity and a higher standard of living to those engaged in it? Second, how can living conditions in the country be improved, particularly housing and the services of the house, so as to bring standards of comfort in rural areas more into conformity with those of the towns? Third, how can the handicap which the small scale of so many village communities imposes upon the organisation of the churches, upon education, on all the welfare services

> and the help, spiritual, moral, and material, which the nation sets out to provide for the countryman and his family, be removed? Lastly, given satisfactory answers to these questions, can anything be done to preserve the amenities of the countryside and the beauty of the rural scene, so that the destruction and the desecration arising from want of thought, from lack of taste, or from the pursuit of profit, which were spreading through the country on an ever-increasing scale in the years before the war, may be brought under control? (pp. 8–9)

Points one and four are properly the concern of other chapters but points two and three are at the heart of this chapter. Both these, it will be seen, were taken on board by planners and administrators after the war with varying success. The record will show that, in broad terms, the matter of rural housing conditions was substantially solved (though with notable exceptions for some elements of rural society). The third point, that of the social and economic problems resulting from the small scale of rural communities, has proved much more intractable.

Orwin's expressed concern for living conditions, services and the diseconomies of small scale was prescient: these themes were certainly to be at the forefront of planners' concern in the second half of the twentieth century. But both themes would be massively influenced and guided by two major controlling forces which he failed to predict. The first was the influence of population growth, a theme which is returned to in chapter 10. Orwin, as with all commentators at the time, was mindful of the debilitating effects of depopulation and saw rural policy as starting from a

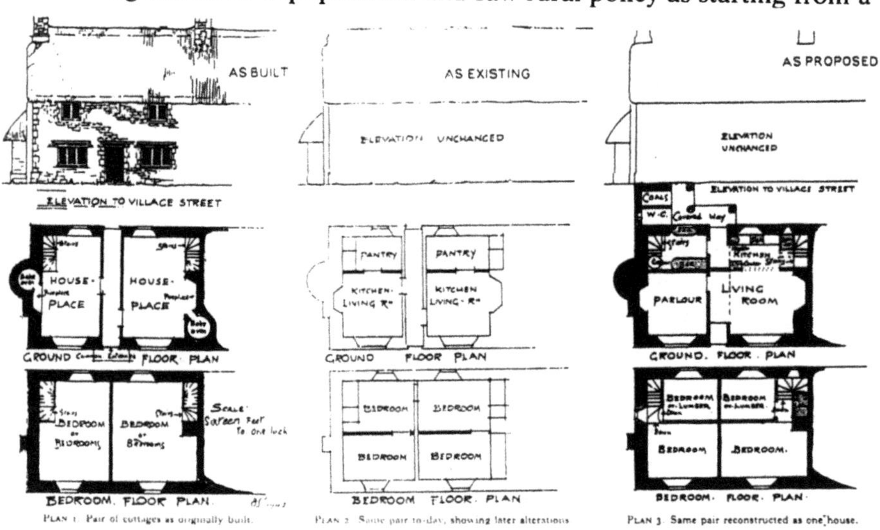

Improvements to rural housing. A suggested improvement to poor quality rural housing taken from *Countryside Planning*, C.S. Orwin's classic study of north Oxfordshire (1944). *Left*: Pair of cottages as originally built. *Centre*: Same pair today, showing later alterations. *Right*: Same pair reconstructed as one house.

need to stem the flow of people to the towns. He could not foresee a future where, for many parts of the countryside, the problems would be associated with substantial population growth. That rural population growth on a massive scale could take place, and yet rural communities could still be faced with the scale and servicing problems which Orwin outlined, remains a paradox which he would undoubtedly have found confusing.

The second force which comes through an analysis of rural community change since 1945 is the political one. The task of planning the countryside seemed, at the end of the Second World War, to be essentially a technical one, untrammelled by political influences. There is a strong feeling in the writings of the 1940s and 1950s (and referred to in chapter 5) of the obviousness and logic of State involvement and action. Rural communities, as indeed agriculture and rural industry, would be subject to objective survey and assessment. Policies would then be formulated based upon measurement and rational thought and then implemented (successfully) in a spirit of cooperation and agreement.

The post-war experience was, in fact, very different. True the general belief in the efficacy of State planning lasted well into the 1960s, but doubts surrounding certain strategic policies surfaced much earlier, particularly where contentious policies such as those for 'key' settlements were concerned. Then pressure groups, gathering a constituency particularly at the local level to fight some hotly-contested issues such as a new housing estate or the threatened loss of a village school, quickly emerged to challenge the simplistic application of the planners' policy. Finally at the grander, national scale the period saw a major turnaround in the whole attitude towards State involvement. While the early years after the War had seen a general consensus as to the role of the State in broad issues of welfare, the powers and effectiveness of bureaucrats (particularly planning bureaucrats) were increasingly questioned by the 1960s. By the 1980s these growing doubts had led to very significant retrenchment in State involvement and its partial replacement by the orthodoxy of the free market and self-help philosophies.

SOCIAL RESTRUCTURING AND COMMUNITY CHANGE

The study of rural communities for the first twenty years after the Second World War was frequently viewed from a position of an antiquated and very formalized social anthropology. The studies of particular rural communities, for example Williams's (1963) study of Ashworthy in Devon (see also Frankenberg, 1966), are justifiably regarded as classics. In retrospect, however, they seem to record the last vestiges of a nineteenth-century community structure but to put them foward as still applicable to the middle of the twentieth century. In their choice of marginal, often upland, locations, they followed the trail of the classical anthropologists bringing to the study of British rural communities an approach akin to the exoticism of studies of New Guinea. Other studies which did not come from this anthropological tradition also tended to portray a static and undynamic rural community. This is well-exemplified in Howard Bracey's study *English Rural Life* (1959) which, just as with Orwin's study noted earlier, provides an excellent overview of the organizations and broad structure of the English village at mid-century, but which with hindsight gives little idea of the massive processes of social and economic change which were beginning to get underway.

Change, and particularly social change brought about by demographic growth and improved transportation, was recognized by these writers but usually as an unwelcome threat to the established patterns they were so keen to describe. It was not until the mid-1960s that a significant change was really discernible. In 1964 Ruth Crichton published her study of Stratfield Mortimer under the title *Commuter's Village* and thus provided a bridge between the old and the new in the sociological understanding of rural communities in the second half of the twentieth century.

But the real mould-breaker in this direction followed the next year when Ray Pahl published his study of commuter settlements in the green belt of Hertfordshire (Pahl, 1965). Though Pahl was soon to leave the study of the rural behind, his work set a tone which pertains today. He did this in two ways. First, and most obviously, he recognized the social reality of twentieth-century rural places, seeing them not as composed of residual groups with just a passing reference to the newcomers, but as *restructuring* communities with the 'new' people in full occupation. Second he made it possible for the study of English rural sociology to benefit from a revival which it desperately needed. It had become a backwater area of study even in America; in Britain it was moribund and all but dead. Pahl showed that the new approaches of the urban sociologists and of the social thinkers could be applied to rural issues.

In a simple but effective way, Pahl presented the new typology of rural communities (Pahl, 1965, 1970). His eight social groupings are still useful as an elementary checklist to aid our understanding of the rural community:

1. large property owners;
2. salaried immigrants with some capital;
3. 'spiralists';
4. those with limited income and little capital;
5. the retired;
6. council house tenants;

7. tied cottagers and other tenants;
8. local tradesmen and owners of small businesses.

The categories are a blend of the old, the new and the altered. Some groups remain apparently unchanged from earlier in the century, but in fact even these have been influenced by modern trends. Thus the large landowners (Group 1) can still show aristocratic lineage and occupation over centuries, but they may also be the *nouveaux riches* with City money or even the faceless insurance company or investment fund. Equally, as was seen in chapter 6, the farmer group remains but often in altered form as agribusinessmen and small farmers with diversified economic interests increasingly replace the more narrowly-focused food producer.

Pahl's studies in the metropolitan fringe were followed by others which also moved away from the upland, marginal and relict rural community to study social change in the 'normal' village of the lowlands where urban pressures were increasingly evident. The south east provided the focus for Ambrose's (1974) study of change over the century 1870 to 1970 in Ringmer, Sussex and also for Connell's study of Surrey villages (1974, 1978). These studies followed Pahl in recording the changed social structure, but, perhaps because their authors were geographers rather than sociologists, they tended to fight shy of much interpretation of the new community dynamics which were developing. Most obviously, these studies seem in retrospect to be curiously unwilling to discuss class relationships and matters of resource competition, whereas Pahl's earlier studies had begun to consider those themes.

The fuller development, not only of the 'social map' of the changing village community but also of the dynamic relationships between social groups, was left to Newby who, after Pahl's start, can probably be credited with the turnaround on rural sociological matters noted above. Newby's studies of farm workers (1977) and subsequently of East Anglian landowners (Newby *et al.*, 1978) concentrated on the relationships between the groupings within the agricultural community. He saw the arrival of commuters and urban migrants not just as the addition of more, and socially different, people to the countryside, but as providing the essential axis around which social relationships were being redefined. The advent of newcomers in large numbers, with different ideas, social background and wealth, forced existing groups to reassess their own position and relationships.

Newby followed Pahl in providing a single idea which was seized on by other workers and by rural social commentators. He contrasted the old 'occupational' village community, where virtually all residents were directly or indirectly involved in farming and where class relationships were closely linked to land ownership, to employment functions (master and man) and to family history, with the new community where the advent of new people without these social anchors reduced the farm-focused groupings to an 'encapsulated' community. While class antagonisms stemming from employment were still to be found, these residual classes found some common cause in their mutual feeling of being invaded and often misunderstood by the new urban groups.

These themes of class conflict have provided a common thread through the significant number of subsequent studies. Useful reviews and assessments can be found in the work of Harper (1989) and Rogers (1993). In some senses it might be argued that the focus of study has now moved completely away from the residual, old-style rural groupings. Thus Cloke and Thrift (1987) have argued that, to all intents and purposes, the 'service' class has so invaded the rural communities of lowland southern England that little or nothing remains that is demonstrably rural. Whatever the truth of this, the revised study of social conflict in rural communities has brought forth a rich body of work which has particularly focused on two areas that have become major issues in the understanding of the post-war countryside.

First these new social groupings and relationships have thrown into question matters of land use, conservation and environmental concern which were previously largely uncontested (Harper, 1993). Put simply, the new people in the rural areas, often well-versed or at least with strong personal views on such matters, have ideas and objectives about how 'their' countryside should be used. Most obviously they see modern farming practices as inimical to the health of both the ecology and the economy of the countryside and they take up a decidedly anti-development stance. Further, because they are articulate and often well-connected to channels of political power, they are prepared and able to act in support of their opinion.

Second, and of more direct concern to this chapter, new people with new wealth and new ideas have precipitated a debate across a wide front which relates to the quality of rural lifestyles. Because rural living has become both possible and attractive for a large number of urban people, the material circumstances of that rural living have become precious. Life in rural communities in the latter part of the twentieth century has become what Fred Hirsch (1977) termed a 'positional' good, having an esteemed value well in excess of the material benefit which it gives. The resources of the rural areas, most obviously housing, soon move out of the reach of the poorer elements in the rural community.

RURAL LIFESTYLES

The task of rural reconstruction outlined by Orwin and others at the end of the Second World War certainly included a social and community element within the major objective of rural economic advance. The improvement in the living standards of the whole rural population was seen in the late 1940s as naturally flowing from two major platforms of post-war rural planning policy: the revival and continued support of the farming industry and the protection of the countryside from urban development. This dual approach would be mutually supportive. Substantial State support for agriculture and the improvement in the health of farm businesses and farmers' incomes would, it was assumed, automatically filter through to the rest of what was still seen essentially as an occupational community. At the same time the planning system would protect the basic agricultural resource from urban incursion and thereby safeguard its economic welfare.

This dual approach, and the supposed benefits to the wider rural community, held fast for a remarkably long time. As chapter 6 explained, the support of agriculture remained a central tenet of rural policy until the mid-1970s and even after this the argument that what was good for farming was good for the whole rural population has been frequently stated.

There was, of course, a third strand to post-war policy for the rural population but not one which necessarily had a specifically rural flavour. This was the development of the national welfare state and its automatic extension to the rural areas. No distinction was made, at least in theory, between the urban and the rural in this regard. Systems of State schooling, of health care and of welfare benefits were seen as available to all, regardless of their location. Beyond the welfare system, other aspects of State support for livelihoods were also seen as generally applicable. Thus in public transport provision, substantial cross-subsidy from profitable urban and interurban services to maintain a basic rural network was taken for granted even up until the 1980s.

The town and country planning system served to bolster this assumption that rural areas should not lag behind the towns. A key element in virtually all the development plans produced after 1947 was a survey of the service provision levels to be found in villages and hamlets. These frequently followed an assessment of the minimum population needed to support key services and a statement of some services thought crucial to the maintenance of the rural community. A good example of this approach can be found in the case of Norfolk (e.g. Green 1971) where the recognition of apparently critical population thresholds was still being attempted twenty years later (Shaw, 1976). This broad approach of focusing, through the land use planning system, upon the maintenance of rural communities by ensuring basic levels of service provision was enshrined in official national policy (MHLG, 1967) and was formalized in many counties in a 'key settlement' policy (see chapter 10).

SERVICES FOR THE RURAL COMMUNITY

The provision of services to rural areas has, in fact, been a constant focus of comment and concern throughout the whole of the second half of the century, going beyond its specific role within key settlement and similar policies. For the first two decades after 1945, there was a tacit assumption that national economic growth, coupled with the directing power of State intervention and the land use planning system, would slowly but surely extend urban-level service standards to all but the most marginal of rural areas. A good example of this attempt to bring urban standards to the rural areas is to be seen in the passing of the Rural Water Supplies and Sewerage Act of 1944 whereby central government could contribute to the often high cost of schemes by the water authority (Parker and Penning-Rowsell, 1981). This measure went far in improving the poor rural living standards which had been highlighted in the Scott Report of 1942. Gradually, however, it became clear that, national and rural prosperity notwithstanding, this automatic modernization of the countryside was simply not happening. By the 1970s the clear focus of comment had become a concern for the 'loss of rural services'. Rural people, not least the 'incomers' from the first waves of counter urbanization, were bemused to observe population *growth* accompanied by continued service *loss*. A report by the Standing Conference of Rural Community Councils (1978), reporting in a very elementary way the loss of key services in rural areas of the South West of England, seemed to touch a national nerve and initiated a large and prolonged period of comment in *The Times*. Further studies followed, emphasizing the trend in all areas, and the level of service provision in rural communities is clearly still seen as a key indicator of prosperity and a starting point for rural policy.

A decade or so on from the 1978 study, a further investigation in a number of rural areas (Clark and Woollett, 1990) reported a mixed picture of service change, though one with decline as a generally common feature. The assumed importance of service provision to rural policy was further emphasized by the publication by the Rural Development Commission of the first national survey of service levels and service change carried out in 1991 (Lievesley and Maynard, 1992).

While the measurement of loss of rural services has certainly served to draw attention to rural problems, it appears in some ways as a backward-looking focus which, by its very historical continuity from immediate post-war planning concerns, has become a less and less useful starting point for policy development. This is not to say, however, that there have not been some significant changes in approach regarding rural services, especially since the 1970s. In the first place there is a clearer recognition by both rural dwellers and by policy-makers that the land use planning system will not of itself provide the policy answers to perceived service provision needs. The role of other sections of public provision, notably in areas such as health and education, are now seen as equally important. In addition the land use planning system has been increasingly regarded, not as the saviour of village infrastructures (by maintaining population thresholds), but rather, by denying new development and growth, as a major contributor to service loss.

Secondly the post-war period has seen the increasing significance of the influence of *national*, as opposed to specifically rural, policies in the area of service provision. Thus, as will be seen below when more particular cases are elaborated, it has been *national* policies for housing or transport, schools or health care which have been the major influences upon rural communities rather than some special set of *rural* circumstances. Rural planning, it might be argued, was for the first twenty years after the war a subset (and indeed often rather an undistinguished subset) of the new discipline of town and country planning. For the twenty or more years that followed it has the growing concern of a wider set of professionals. The special problems of rural areas, flowing from their inaccessibility, have to some extent been recognized in these separate policy areas by a continuation of broad policies of uniform charging and also by government recognizing a so-called 'sparsity factor' whereby the financing of the more rural local authorities has reflected the extra cost of service provision. But both these aspects of special treatment are under threat in the 1990s, on the one hand from the privatization of many of the service provision agencies and on the other from financial cutbacks in a recession-bound economy.

From 1979 the Thatcher government instituted a process of privatization which had far-reaching influences upon rural communities (Cloke, 1992). Major areas of service provision, notably water resources, transportation and telecommunications, were moved from the 1980s into private corporations. Arguments from the rural lobby that this would lead to disadvantage for rural dwellers if charges for services were to reflect the perceived higher cost of rural supply, were met by specific requirements of the new companies that uniform policies should continue and that rural consumers would continue to pay the same as their urban counterparts. These requirements will undoubtedly come under increasing pressure during the 1990s as privatized service providers become less willing to cross-subsidize the smaller rural market. The economic logic underlying the key settlement approach (Cloke, 1979, 1983; chapter 10), stands in direct opposition to a post-war tradition of welfare provision. It remains to be seen which approach will prevail.

While the debate on rural services, and the policy response, has ranged widely over many types of provision, it is fair to argue that three particular elements have tended to elicit most attention. Transport, education and housing have frequently been the especial focus of concern and all three areas exhibit the privatization trend noted above.

Rural Transport

Rural transport problems emerged early in the post-war rural debate as a key issue. In the immediate post-war period the necessity of maintaining and even enhancing a rural public transport system, based mainly on bus services, was seen as self-evident. The government-funded film based on Orwin's survey, referred to at the outset of this chapter, made the point clear: the problem was one of overcrowded buses and insufficient services to meet the growing need from the rural population.

Post-war prosperity, and in particular the rise of the private motor car soon changed the nature of the problem. By the mid-1950s private car transport had overtaken the public bus service as the main form of rural transportation. As passenger numbers fell, so the National Bus Company, which had been created by the Transport Act 1968, responded by increasing the element of cross-subsidy from more profitable urban services and, eventually, by cutting back on rural routes. The process of decline after about 1955 was inexorable and it has continued to the present day. Official recognition of the problem came with the report of the Jack Committee (1961) which attempted an analysis of the rural bus problem and to suggest solutions.

Until the end of the 1960s the basic policy approach was predicated upon a view that a rural public transport infrastructure should be maintained if at all possible. Bus services were seen as providing the main mesh of this infrastructure, with rail services providing a coarser network linking larger rural places with towns and cities. There was, in retrospect, a curious inability to recognize the fact that rural rail services had been gradually declining since the inter-war years at least. Thus the proposals in the Beeching Report (1963) to close many rural lines was seen by many as a sea-change, though it was more properly simply a formal consolidation of an ongoing process.

The policy approach to the emerging rural transport problem has been broadly threefold. First, the demonstrable withering of a conventional comprehensive transport network was met with a policy of experimentation and alternative approaches. A series of sixteen official experiments in particular regions including Devon, North Yorkshire and Dyfed was instituted in the 1970s (Balcombe, 1980). In practice these experiments seem to have remained just that and have not opened up a comprehensive new approach to rural transport.

Second, and linked to some extent to this experimentation particularly since the early 1980s, there has been increasing expectation that the voluntary sector will provide transport in rural areas where the public sector response is deficient, particularly for specific services such as hospital visiting (Banister and Norton, 1988). In part this has been little more than a tacit recognition of the substantial role played by volunteers within rural communities over many years but it has also been formalized as part of official policy. Thus from the mid-1980s, the Rural Development Commission has operated a Rural Transport Development Fund to encourage innovative rural transportation schemes. Such provision has also been a common activity of the county-based Rural Community Councils (see below).

The third strand of policy, as in other areas of service provision, has reflected the major change in political ideology which was identified with the Conservative government elected in 1979. The 1980s saw the progressive denationalization and deregulation of bus services, with express coach and stage-carriage services privatized in 1980 and other routes, more significantly for rural areas, following in 1985. The argument was that increased competition and deregulation would cut costs and so help to improve the level of service to rural communities. The results, at least by the early 1990s, were at best varied but with every indication that few areas actually saw an increase in provision while many continued to register a loss (Bell and Cloke, 1990). For those rural areas still served by a rail transport system, the prospective privatization of this service in the 1990s will result, many would argue, in a reprise of the experience with road transport.

Rural Schools

Rural education, predominantly at the primary level, has been at least as important in the minds of rural communities as rural transport. The post-war period started in much the same way as with other rural services, with the consolidation of a national approach to education in R.A. Butler's Education Act of 1944. But the reality of maintaining schools in the countryside impacted upon the newly-created local education authorities (the County Councils) if anything quicker than in the case of transport. While the pace of change varied over time and from county to county, village schools closed at a rate of about one per week.

In part this was due, as with other service areas, simply to the economics of maintaining small schools where costs were high and where buildings dated from the previous century (Stockford, 1978). But other factors were also at work. Some were apparently political as when changes of local government brought to power those local authorities (generally Labour-controlled) which favoured the urban areas and saw less need to bolster the rural school network. Others stemmed from educational philosophy, most notably when the Plowden Report (1967) on primary education in England – and its equivalent in Wales, the Gittins Report (1967) – seemed to favour larger schools (above 50 pupils) on the grounds of the quality and range of the education which they could provide. And finally the declining birth rate, which fell nationally after 1964, was also a significant factor especially when bolstered in many rural areas by the incursion of childless middle-aged and elderly newcomers.

Throughout the second half of the twentieth century it is not really possible to discern a distinctively *rural* education policy. Indeed education is probably the most obvious area where the necessity for national (i.e. predominantly urban) standards has been assumed by government and where, therefore, rural communities are generally forced to accommodate themselves to a national, centrally-directed pattern (Tricker and Mills, 1987). Certainly the attitude which existed in many rural areas until the end of the 1930s, that rural children were somehow 'different' and that rural education should accordingly follow a different pattern from what went on in towns, largely died with the Second World War. It is, nonetheless, salutary to remember that until then it was common in some rural areas for farmers to require children to forsake schooling at critical times in the farming year when they were needed for work in the fields (Ward, 1988).

The general absence of a national policy approach to rural education was not, however, matched by the public response to the continued loss of rural primary schools. The threatened loss of the local school provided a constant source of local protest, action group formation and political lobbying at all levels. Paradoxically, while governments of all persuasions have been at pains to emphasize uniform national standards, rural people have often argued for their school on the specific grounds of rural education being different, with personal care, supposedly higher standards and the school's role within the community being cited in defence. In these arguments local groups have been aided by a number of national pressure groups, led by committed and energetic campaigners, notably the National Association for the Support of Small Schools.

The frequently argued role of the village school in rural community life prompted an official study by government in 1981 (JURUE, 1981; Tricker, 1983) which tried to ascertain whether the loss of the school really did influence community life adversely. The results of the study were often vague and seemed to suggest that, after an immediate crisis following closure, the absence of a school was largely unrelated to community well-being. This indeterminate result has not, of course, hindered the use of the 'community' argument by enthusiastic supporters of rural schools.

Housing the Rural Population

The third specific area which may be considered in more detail, rural housing, has arguably become the major rural service issue of the last quarter of the century. Rural housing, however, cannot sensibly be seen within the structure of continuous service loss which can be applied in other areas.

From the end of the Second World War until perhaps the end of the 1950s, the rural housing debate centred on the traditional focus of poor housing conditions (Rogers, 1976). Until then, the quality of rural housing was seen as a key problem for rural development; the Scott Report of 1942 had emphasized it and in the immediate post-war years local authorities were encouraged to improve housing quality as a major contribution to rural reconstruction. In contrast to the pre-war period, no specific *housing* legislation focused solely on the rural areas. Rural housing conditions were improved through the medium of national legislation providing grants and setting standards for new housing. By the 1960s the condition of rural housing as a prime area of planning concern had largely faded from view. A residual problem of unfitness and lack of amenities, affecting perhaps 5 per cent of the rural housing stock, remained (as indeed it still does), but the pre-war horrors of rural living as caricatured by Stella Gibbons in her rural pastiche *Cold Comfort Farm* (1932) were things of the past.

New housing in villages. Stratfield Mortimer in Berkshire, the focus of Ruth Crichton's book *Commuters' Village* (1964), an early study of social change in rural areas. (By courtesy of the Rural History Centre, University of Reading)

For a brief while in the late 1960s and early 1970s the focus of attention moved to the growth of second home ownership (Downing and Dower, 1972). In its own way this new area of apparent concern was both symptomatic of the times and also a curtain-raiser to more deep-seated housing issues which would follow. The 'second home issue' was seen initially, not as an issue about rural housing, but rather as an aspect of the newly-emerging field of countryside recreation. The first research in this area (Bielckus, Rogers and Wibberley, 1972) was carried out for the Countryside Commission as one of the first investigations following its formation under the provisions of the 1968 Countryside Act. As the 1970s progressed, however, while there was still interest in such issues as the contribution of second homes to local rural employment (e.g. Jacobs, 1972), the focus was rapidly changing towards issues of rural housing equity and access rather than of recreational provision. The early research had, in fact, begun to raise the question of adverse effects of competition by urban outsiders for a limited rural housing stock. By the end of the decade the disadvantage of many local people in this competition had become a major concern of planners in rural areas.

It is significant that this question of housing competition and perceived local need was raised by planners, as opposed to housing managers, and was especially to be found in highly attractive areas, notably in the National Parks. It was naturally these areas of high landscape quality, made all the more accessible by the developing motorway network, which increasingly attracted second home owners. Moreover it was arguable that the restrictive planning system was in itself at least partially to blame for high rural house prices and competition from outsiders. The issue effectively came to a head at the end of the 1970s in the Lake District, an area where the initial second homes research had shown at the end of the 1960s that antagonisms were already present and where, in retrospect, it is possible to see the specific second home issue being transferred into a much broader concern for 'homes for locals' during the 1980s.

The conflict faced by the Lake District National Park authorities was in essence simply an extreme example of the problem faced in all rural areas in the post-war years: how to safeguard the welfare of the local community, particularly with regard to improved housing and housing choice, while at the same time protecting the countryside from excessive development. The answer appeared to lie in the restriction of new housing development to that intended for local people only. The policy, introduced first in 1977, was at first unpopular with the Department of the Environment such that it was deleted from the 1981 Draft Structure Plan. The ideological swing after 1979, highlighted by the election of the Thatcher government, meant that the policy was seen as undue public interference in the housing market and an unwelcome extension of planning powers to consider questions of who lived in a house rather than strict matters of land use change. The particular policy measure in the Lake District was therefore withdrawn in 1984.

The apparent failure of the Lake District to maintain its 'locals only' policy at the outset of the 1980s was, in retrospect, premature. Official disapproval of such intervention on the behalf of 'local' people gradually disappeared until by the end of the decade central government was even emphasizing the need to supply housing for local people. Planning authorities were allowed to recognize exceptional circumstances where 'local housing' should be built where 'general needs' housing would be refused (DoE, 1991). By the time that the Cumbria and Lake District Joint Structure Plan was reviewed in the early 1990s, a 'locals only' policy had been reinstated. At the Examination in Public, while there was some opposition from development interests, the policy was effectively upheld by the Panel who argued that they 'were disposed to give the Structure Plan authorities the benefit of any doubt [they] might have in deciding that environmental consideration should have a paramountcy in questions of development in the very heart of the Lake District' (Cumbria and Lake District Joint Structure Plan, 1993).

The recognition by policy makers during the 1980s of a particular category of people ('local' people) in fact marked a most significant change in rural planning policy during the second half of the century. For many years from 1947 planning policy was concerned essentially with physical land use change. The tacit assumption and logic was that proper planning focusing on controlled development and good design would of itself give rise to good social outcomes. It was not the planners' job to define categories of people and thereby get involved explicitly in social engineering. With greater impact from the social sciences during the 1960s and 1970s, land use planning had increasingly been seen in a social and economic context. The first wave of Structure Plans in the 1970s was quite explicit in this direction. However the Department of the Environment trimmed this contextual enthusiasm and by the 1980s statutory development plans had contracted much more to their core concerns.

The growth of concern for local housing need during the 1980s encouraged policies which specifically targeted local rural people. A few commentators provided strong criticism of such policies. At an early stage Shucksmith (1981, 1990) had studied the Lake District case and had shown how the new 'locals only' policy probably *increased* prices to local people rather than making housing more affordable. Equally Rogers (1985) argued on a number of grounds that such policies were ill-founded as a basis for policy. But these criticisms were not heeded because the mood of the times suited local need policies. Planners and politicians felt that they must be seen to be acting, while the public view, especially in the rural areas, supported them. Preferential policies favouring local people appeared as the embodiment of the real rural community and, as some commentators noted, 'locals only' policies could be seen as the acceptable face of otherwise strict development control.

Paradoxically, as the Department of the Environment was in the 1980s gradually espousing the need to provide housing for local rural people, a second main strand of housing policy was arguably having a directly contrary effect. Proposals to sell council houses to their tenants had been an important factor behind the Conservative election victory of 1979. The Housing Act 1980 provided the legislative basis for this promise whereby substantial discounts were available to long-standing tenants. Concern amongst rural lobby groups, including the National Farmers' Union, regarding the likely impact in the countryside led to a token concession to recognize a small number of designated 'rural areas' (mainly in the National Parks) where some controls upon the resale of former council houses would be possible (Phillips and Williams, 1983). In practice these powers have rarely been used and sales of council houses in rural areas have pushed ahead if anything at a greater rate than in the towns and cities.

The provision of housing in rural areas is, of course, just one example of a major political sea change which occurred after 1979. Prior to this the post-war consensus as to the role of the local state in the provision of welfare was maintained by successive governments. After 1979 the role of the local authorities was severely cut back and many of the tasks which they previously performed were effectively given to either the private or to the voluntary sectors. Private house building certainly increased substantially relative to the public sector, but economic crises and a major depression towards the end of the decade inevitably limited its absolute growth. The need to replace the role of local councils as the providers of social housing was barely recognized by government in the early 1980s when the supposition was that the private sector would provide. The gradual rise in significance of the 'local need' issue noted above meant that a more positive role for the voluntary housing sector was gradually, if belatedly, developed. By the end of the 1980s housing associations were beginning to contribute many more 'affordable' homes in rural areas though these numbers appeared small in comparison to some estimates of rural need during the 1990s which were estimated by one authority to be as high as 370,000 (Clark, 1990, 1992).

The broad post-war trend of concern in rural housing moving from housing quality to housing access and local need has tended to overshadow one further element of the issue which has had a long history in rural policy during the twentieth century – the agricultural tied cottage. Traditionally the tied house was seen as an essential component of an efficient farming industry. Though its abolition was an aim of the farm workers' union from as early as 1909 and became part of official Labour Party policy after the Second World War, the power of the farming lobby within the political economy of the rural areas and at a national level meant that outright abolition was never really likely. The post-war world saw the continuation of class antagonisms in agriculture focusing on the tied cottage (Rogers, 1976) only slightly ameliorated by new council house building in some rural areas in the 1950s and 1960s. Eventually an attempt was made in the mid-1970s to solve the problem. Under the provisions of the Rent (Agriculture) Act 1976, tied cottagers were given the same rights as other tenants and a system of Agricultural Dwelling House Advisory Committees was created to adjudicate in disputed cases. Where the needs of the farmer to gain possession of a house were proved, local authorities were expected to find council accommodation.

Despite concerns during the 1980s that the sale of council houses would effectively subvert the workings of the 1976 legislation, there seems little evidence that this in fact happened (Rogers, 1992). While still an issue in the minds of the farm workers' union, the pre-1976 antagonisms seem to have disappeared. The 1976 legislation has now effectively been replaced for new tenancies by the Housing Act 1988.

DEPRIVATION AND POVERTY IN THE COUNTRYSIDE

The post-war recognition of problems in rural communities has tended to be issue-based, focusing on particular themes such as housing or service loss. Moreover the problems of rural lifestyle have tended to be seen as having their origins in the particular physical characteristics of the countryside, especially distance and inaccessibility (Moseley, 1979; Lowe, Bradley and Wright, 1986). As was noted earlier, this view led inevitably in the decades after 1947 to the assumption that the proper implementation of the planning system would, through the provision of services, fully address any supposed deprivation in the countryside. Such planned provision of services would be bolstered by a welfare state which, by being applicable to both rural and urban residents, would safeguard standards of housing quality, education, medical care and the like. The more precise question of the rural poor would also be addressed by national policies founded on the principle of universal benefits and income support and a strong domestic farming industry.

By the end of the 1970s there was a growing realization that this post-war optimism was not justified. Partly spurred on by attempts by the shire counties to attract more funding from central government (Association of County Councils, 1979) and by a government-funded study of rural deprivation (McLaughlin, 1986), the late 1970s and early 1980s saw a major debate on the significance of deprivation and poverty in what on the surface appeared a healthy and prosperous countryside (e.g. Shaw, 1979). Research suggested that perhaps one in five of rural residents were living with incomes which could be considered as below the 'poverty level'. While these figures were disputed at the time, they did provide a useful basis during the 1980s for proponents of particular policies to argue their case. At first government was loath to accept these arguments, questioning the statistical validity of the research and refusing official publication but also arguing that the new liberalization of many rural policies in the 1980s would in any case provide ready solutions to problems. By the end of the decade, however, the existence of a substantial underclass in rural areas was tacitly accepted. Indeed a further officially-sponsored research study was funded in the late 1980s, partly initiated by the work of the Archbishops' Commission on Rural Areas (ACORA, 1990). The eventual publication of this second study (Cloke *et al.*, 1994), while certainly better-founded statistically, confirmed the findings of a decade earlier. In the 1990s there is a greater recognition that the 'chocolate box' image of the English countryside often hides human problems which are every bit as real as their urban equivalent (Derounian, 1993).

RURAL GOVERNANCE

For the first three decades after the Second World War, the countryside was subject to a local government system which in broad outline had been created at the end of the nineteenth century (Young, 1989). This system recognized the particular nature of the rural areas in two ways. First there was a category of Rural District Councils fitting underneath a higher County Council level and paralleling an equivalent Urban District category. Secondly a system of parishes, created by the Local Government Act of 1894, was seen as reflecting very local needs at the village level (Poole and Keith-Lucas, 1994).

The Rural District Councils provided local services such as housing and refuse collection while the County Councils, along with the provision of police and fire protection, had responsibility for education and for the newly-developed planning system. At the parish level, powers were strictly limited with many councils able or willing to do little more than oversee minor matters such as street lighting and rights of way. Antagonism between the various levels of government inevitably existed but to a certain extent they were ameliorated by informal class connections within local government. Thus, as Newby (1979) showed as late as the 1970s, the local government system was disproportionately dominated by landowners and farmers who inevitably brought their particular views to bear on issues and policies. Moreover there was significant linkage between the levels of local government with the same people representing an area at two or even three levels.

This system of local government was modified in 1974 by the provisions of the Local Government Act 1972. The old Rural and Urban District Councils disappeared and were replaced by single authorities (District Councils) which in general were larger and which linked an urban centre with its rural hinterland. At the time there was opposition from rural dwellers to the new pattern since they feared that rural perspectives would be overwhelmed by the greater influence of the town. Partly as a sop to this view, parish councils were given some extra powers, especially to comment upon all planning applications in their area. In practice, while this has clearly worked well in some Districts, it has not in others either because of parochial lethargy or because the District has been unwilling to share its powers in this way. Moreover, while there are many examples of vibrant and active parish councils, many remain apparently moribund and in effect have by their inaction condoned a domination of urban power over local determination (Collingridge, 1987). A major review of the whole local government system commenced in the early 1990s and holds out some prospect of enhanced powers for parish councils but it seems likely that the pattern of such influence will continue to vary widely over the country as in the past.

Over and above formal structural changes consequent upon legislative reform, a major development of key significance for local governments relates directly back to the demographic and social changes which were noted earlier in this chapter. The counterurbanization trend and the invasion of the countryside by the service class have inevitably impacted upon local government. The dominance of farmers and landowners on local councils has undoubtedly been waning since the middle of the 1960s as newcomers have moved to rural areas and have been elected at all levels of local government. The decline of the farming influence in their areas has also arguably been compounded by a farming crisis in the 1980s which has often curtailed the role of the farmer in local politics as his business pressures have become more demanding.

The formal structure of local government outlined above is, of course, only one aspect of rural governance (Cloke and Little, 1990). Of at least equal importance is the network of non-governmental organizations and voluntary groupings which are important in managing rural areas. They extend from quite formalized institutions such as village hall committees and school governing bodies to a welter of local and national voluntary organizations and pressure groups particularly concerned with environmental protection.

Since the early 1980s the role of these non-governmental groupings has been especially encouraged by politicians keen to reduce the power of the central state and its bureaucracies. Thus local influence, especially of schools by way of management and budgetary powers and eventually by opting out of local government control, has undoubtedly increased. But a second influence in this direction has been at least as important. Just as it was noted above that the new rural middle classes have engaged in formal local government, so have they increasingly involved themselves in the informal governance of rural areas. Sometimes, as with community development (see below), this may have apparently altruistic purposes; at other times the purpose is unequivocally one of self-interest. Local preservation and protection societies, sometimes *ad hoc* in their origins and operation, sometimes linked federally to national groupings such as the Council for the Protection of Rural England, have been ever-vigilant in safeguarding rural amenity and limiting development (Lowe and Goyder, 1983; Buller and Lowe, 1982). While ostensibly those groups attempt to gain their ends by publicly representing 'local opinion', it is clear also that they are often quite closely linked with the formal planning process. A study by Buchanan (1982) of the structure planning process in Suffolk made clear the important role in policy formulation which had effectively been given to the Suffolk Preservation Society by both officers and members of the local authority.

RURAL COMMUNITY DEVELOPMENT

Recognition of the increasing role of voluntary organizations in the area of landscape protection links across to a final element of rural community structure in the second half of the century. Unlike some other countries, it is quite difficult to point to formal policies of rural community development in Britain. The main reason for this can in part be found in the origins of British rural community development in the years after the First World War. The responsibility for community development was from the outset passed by the government's rural development agency, the Development Commission, to county-based voluntary groups known as Rural Community Councils (Brasnett, 1969). These organizations often operated very effectively during the inter-war period, and mid-century studies of rural life, such as that by Bracey (1959), were clear in their judgement as to the benefit which they gave to rural social life.

Until the 1970s the Rural Community Councils, still recognized as providing the formal structure for rural community development, continued in broadly their pre-war direction. Though there were notable exceptions, some became fossilized and backward in their thinking and working, content to believe in a largely self-sufficient and self-helping ideal rural community and unwilling to recognize emerging social problems and tensions. At best many simply continued a work pattern more characterized by tradition and worthiness than by active involvement with rapidly changing rural communities. At worst some Councils employed staff whose idea of community development was based more on the tenets of 1930s colonialism (Wright, 1990, 1992) than on the professionalism that could be found in urban-based community work.

A significant change in this rather moribund picture happened in the 1970s. Government recognized a potential for active involvement in the Rural Community Councils and from 1973 funding was made available for field officers whose role was to engage directly with rural residents. At the same time there was developing a growing interest in the problems of rural areas which was matched by a supply of young graduates from the universities who were eager to take on this new role. While some Councils were laggard in adopting new ways of working, there is no doubt that a major revolution took place in rural community development at the time. This change was to proceed further in the 1980s, culminating in the creation of the organization ACRE (Action with Communities in Rural England) to act as a national association of Rural Community Councils (Rogers, 1987).

The tradition of self-help in British rural community development which stretches back to the 1920s is still very much at the heart of the policy approach (Francis and Henderson, 1992). There is little doubt that it has progressed much since the years after the Second World War and it is today often characterized by challenging and innovative ways of working (Henderson and Francis, 1993). For some, however, it remains an approach which is seen as severely limited, linked too closely with the rural class structure and outdated ideas of charity. To those critics it appears unlikely to be able to deliver radical grass-roots policies and it can be seen as allowing central government to abrogate responsibility for a thorough-going policy of community development in rural areas (McLaughlin, 1987). Whatever the view adopted, the commitment to self-help and voluntary action which is at the heart of the approach appears very much in line with the image of the rural community held in the minds of many newcomers to rural areas.

CONCLUSION

What of the problems which Orwin itemized in 1943 with which this chapter started? The two issues which very particularly focused upon the rural community related to the quality of living conditions, especially housing, and the problem of small scale which handicapped the efficient provision of village services.

Regarding the first, it is fair to argue that a great deal has improved over the last half century. The concern that rural conditions lagged seriously behind those in the town has in many ways been reversed. While remaining problems of deprivation and poverty certainly exist, and probably affect a larger proportion of the rural population than is commonly supposed, the policy focus in the post-war years has more often centred upon conditions in the inner cities. The problems of small scale substantially remain, however, and provide a major handicap to village servicing, affecting both public and private services by way of extra costs and declining levels of provision. Orwin's assumption was that these handicaps affected nearly all in the village community and as such were a proper focus for public policy. Five decades on, however, it is clear that only a minority of this population are seriously affected by the continued problem of scale. For many of the remaining rural residents, private wealth effectively insulates them from such problems so that the countryside has become not a place from which people wish to escape but rather one which attracts more and more each year.

10

RURAL LAND USE PLANNING: POST-WAR PURSUIT OF OBJECTIVES

Countryside planning might be said to date from 1909, when the Development Commission was set up, with terms of reference to support 'any scheme which may be calculated to benefit, directly or indirectly, the rural economy of Great Britain' (Wibberley, 1982). Two million pounds were set aside for its operations from a new tax on the consumption of alcohol, particularly whisky. The then Chancellor, Lloyd George, was influenced by the poverty of the rural economy he had seen during his boyhood in North Wales and later during his political career as he journeyed throughout England and Wales. The Commission proceeded to deal with a variety of socio-economic questions, though its best known agency, the Council for Small Industries in Rural Areas was not set up until 1968, and many of its rural concerns either became the province of other agencies (agricultural research was taken over by the Agriculture and Food Research Council for example) or were absorbed by the growing statutory procedures for town and country planning.

The Town and Country Planning Act, 1932, beginning life as a Rural Amenities Bill, was the first planning legislation to embrace the word 'country' in its title, its provisions no longer restricted to urban areas, as had been the case with Town Planning Acts in 1909 and 1919. In fact County Councils had already begun to display an increasing interest in town planning matters, and the Local Government Act, 1929 extended to County Councils in England and Wales the right to share in the preparation and administration of joint town planning schemes. The legislation of 1932 enabled schemes to be prepared for any land, whether urban or rural. It was not mandatory (the 1919 Act had been, though in practice its take-up was limited); in any case liability on local councils for payment of compensation to aggrieved developers was an obstacle to the widespread use of the powers. By 1942 only 5 per cent of England and 1 per cent of Wales were actually subject to operative town planning schemes, though considerably more was subject to interim development control (Cherry, 1974).

A fundamental change came with the Town and Country Planning Act, 1947. This required the preparation of Development Plans for all boroughs and counties and instituted control over most types of development, defined in the Act as 'the carrying out of building, engineering, mining or other operations, in, on, over or under land, or the making of any material change in the use of any buildings or other land'. But, as noted in chapter 6, the Act specifically excluded the use of any land for agriculture, forestry or building associated with such use from the meaning ascribed to the word 'development'. This exclusion provides a convenient starting point for the present chapter, because so much was to rest on the assumption that post-war attention would be directed to the planning of urban areas and that the task in the countryside would be to protect rural settlements from building encroachment and to safeguard agricultural land. The Town and Country Planning Act and the Agriculture Act, both reaching the statue book in 1947, had complementary objectives and they imposed a strong, strategic hold over land planning in rural areas for many years (indeed in large measure up to the present day).

In this chapter three aspects of post-war rural planning are examined: settlement planning, village design and green belts. They became central to the strategic principles underpinning the location of development, the separation of town from country and quality of rural life. Finally the key elements of land use planning for the countryside are summarized, tracing continuities in an evolving agenda.

POPULATION DISTRIBUTION AND SETTLEMENT PLANNING

The notion of planning was underpinned by the belief that through a rational utilization of the country's resources, greater efficiency and effectiveness would ensue from State control than from a wasteful private market system (see chapter 1). When applied to rural areas this was no less true and it became the task to iron out all manner of imperfections and inconsistencies in the market from land use to spatial patterns of distribution. This determination would apply to the question of the location of population and the spread of settlements.

England and Wales, by virtue of both a relatively settled history extending over many centuries, and a physical geography of great variety, had a population spread based on a multiplicity of small settlements – perhaps 17,000 according to calculations made by Best and Rogers (1973). Such a diffuse network implies economic and social costs in servicing the settlements, only overcome by the development of hierarchical functional arrangements. Such a hierarchy was provided in the past by a network of market towns of varying sizes and functions. This system could be expected to work reasonably well in a purely market-distributive sense while relatively uniform population distributions obtained and were reasonably constant over time. But the twentieth century threw up many distortions of this pattern, including sustained outmigration from certain rural areas, some new population growth in residential fringes, a breakdown of isolation in some districts and a heightening of remoteness in others.

The enthusiasm for the total planning of the countryside in the 1940s offered new opportunities for returning to a greater evenness and rationality in population spread and for considering alternative distribution patterns. In the Netherlands the reclaiming of the Dutch polders had shown that new village arrangements could be devised and in the USA the Tennessee Valley Authority had suggested the ways in which an old landscape could be restructured. In the new commitment to planning, even an old cultural landscape with a settlement pattern dating from Anglo-Saxon times (with precursors in many instances) could be looked at afresh, and in their search for models the planners borrowed freely from other disciplines, notably geography, economics and sociology.

In the 1930s the German geographer Walther Christaller (1933) developed a theoretical explanation of an 'ideal' settlement pattern in which a hexagonal service area surrounding each settlement would allow hierarchies to develop, whereby they could be ranked according to population size and levels of service provision – hamlets, villages and towns in a 'central place' system. British geographers found this spatial theory attractive: R.E. Dickinson (1932) had already pointed to a nucleated rural settlement pattern in East Anglia and Arthur Smailes (1944) attempted to sketch hinterlands (or 'urban fields') of central places in England and Wales. Later, work by H.E. Bracey (1952), concerning Wiltshire and subsequently (1956) in six counties in southern England, found support for a central place model in rural settlements. This enabled him to construct a typology of English central villages which categorized them by the number of shops they possessed and the spheres of influence which could be drawn on the basis of mobile services. Cristaller's work led to one unpleasant aftermath in that his rational pattern of distribution was used by the Nazis as the basis for new German settlement planning in a reconstituted Polish landscape. But in Britain the upshot was more benign: the recognition of the relationships between service centres and their hinterlands, and the deliberate promotion of rural service centres in post-war development. Rural settlement policies based on this prescription readily found their way into the new Development Plans prepared after 1947 (Woodruffe, 1976).

There was some support for the approach, particularly in the 1960s, from directions other than central place theory, though they were related to it (Cloke, 1983). The first was the work undertaken on service threshold analysis. The idea of a threshold as an indicator of the minimum population necessary to support a particular service, facility or activity was not new, but R.J. Green's (1971) research in rural Norfolk suggested a variety of threshold populations for a range of public services such as home nurse, primary school, medical practice, health visitor and secondary school. The size of some of these thresholds suggested the need for a concentration of resources and investment in an increased size of settlements – rural centres with almost urban-scale expectations.

The influence of service thresholds later diminished, though in the early 1990s the Rural Development Commission in a series of surveys of rural services once more pointed to their importance. However, in the meantime work from another direction tended to emphasize even further the need to identify important centres in any settlement pattern. Studies on economies of scale once again suggested larger points of concentration, the argument being, for example, that one large school was a more economic form of provision than several smaller ones, or that diffusion of retail outlets was less viable than a smaller number of larger shops. In the 1970s this fashionable argument pointed to the high cost of providing public services in rural areas with scattered populations.

These three approaches (central place theory, service threshold analysis and economies of scale) came together in the concept of growth centres, whereby rural resources might be built up in the form of ideal villages. In Cambridgeshire, under the influence of the Chief Education Officer Henry Morris, village colleges had been introduced as early as the 1920s. As secondary schools by day, and centres for adult education and social affairs at other times, they functioned as a central point, serving a wider catchment area. This particular scheme seemed to hold out much promise for an approach to rural planning generally.

The first wave of Development Plans was heavily influenced from two perspectives as far as rural policies were concerned. One was the preoccupation with conservation, landscape protection and the separation of town from country; the other was the need to concentrate services and resources within the rural settlement pattern. In practice the two were mutually supportive and county policies followed largely similar lines in three main respects. The first was to promote 'key settlements' (Cloke, 1979) in which there would be a concentration of housing development, employment, services and social facilities. The second allowed for planned decline whereby public support for certain small villages might be terminated, and the population regrouped. The third sought to classify villages according to environmental quality (outmoded mining settlements would rank low) and service capacity (small size, isolation and remoteness would imply gross diseconomies), in order to direct new development to more appropriate locations.

There was little experience on which to base early policies and a rather static view of village development was at first in evidence. There could be few reliable indications as to how village and small town growth might proceed and how population movements might get under way. In the circumstances a good deal of internal flexibility was built in to the provisions of Development Plans in so far as key settlements and classifications were concerned; by identifying a large number of points of concentration, it was a case of hedging bets. The real problem came with policies for planned decline, the best example afforded by Durham County Council. In this part of the north-east a spread of nineteenth-century mining villages had established a settlement pattern which twentieth-century pit closures rendered obsolete, and in the county's classification of villages no less than 114 fell in the notorious Category D which specified those to be allowed to die (Cloke, 1983). Some of the smaller settlements were in fact demolished as the existing houses became uninhabitable, but the policy ran into a barrage of criticism from public opposition and was subsequently modified to a point of withdrawal.

The first generation of County Development Plans was submitted to the Ministry of Housing and Local Government for approval by the mid-1950s. Their revision was in hand during the early and mid-1960s – nearly 20 years after the first rational approach to population distribution planning was mooted. The results of intended policies varied considerably, but the evidence is that while resource concentration in selected settlements was pursued, the difficulties of combatting allegations of discrimination against non-selected villages meant that the policies were not applied rigidly. There was sufficient flexibility in the system to permit private housing to be built to a much wider geographical spread than perhaps intended; indeed in some counties more housing was built in non-selected villages than in those designated for growth (for example Devon between 1965 and 1975), as calculated by Cloke (1983).

The introduction of Structure Plans after 1971 resulted in no great departure from the past in rural settlement policies; there was a general adherence to a concentration strategy accompanied by adjustment to local conditions and changing circumstances. Progressively, local village infilling has taken place and other small scale developments have occurred to ease local housing problems. The pace of change may have been slow and consequently undramatic in immediate impact, but collectively over time the result has been surprising. The fact is that over a 40 to 50 year period the amount of new development in the countryside is greater than could ever have been foreseen. The Council for the Protection of Rural England (1992) have claimed that 705,000 hectares of rural land were urbanized in the 45 years 1945 to 1990. This is about 6 per cent of the country, equivalent to the combined area of Greater London, Berkshire, Hertfordshire and Oxfordshire. In the West Midlands alone an area of farmland twice the size of Birmingham has disappeared since the end of the war.

This situation is a consequence of a greater demand to live in non-urban areas (a rural setting is a preferred location for housing) and a greater ability to meet that demand because of the ease of personal mobility through car ownership. As was seen in chapter 9, the post-war experience has been that different people have chosen to live in rural housing stock. The assumption at the beginning of the post-war period, that the planning task was simply a case of building an appropriate number of houses for a relatively stable population size, in certain carefully selected locations, could not have been further from the mark. Rural settlement planning (as also with urban planning) has been obliged to adjust to the steady decrease since the 1960s in the average size of households and their increasing number (household formation rates therefore rising more rapidly than the population), particularly of one-person households, both for the young and the elderly. Planning, of necessity, has had to be a case of adjustment to changing circumstances, ideal models soon overtaken by events; difficulties arose where plans, policies and programmes, prepared to accord with past assumptions, were not modified fast enough to meet new realities.

One important change to affect the whole of the post-war period has concerned demography. The population of Britain between the wars had been relatively static, by the 1930s the total number little more than replacing itself. Had there been a census in 1941 the country's population would have been shown to be only marginally above that of 1931. On the assumption that attitudes to child rearing would not change significantly, planning for the post-war world could proceed on the basis of little change in numbers. But a host of factors induced considerable departures from notions of a static future. An immediate post-war baby boom was followed by a sharp increase in the national population in the 1960s, which fell away in the 1970s. Average annual increases of 0.5–0.6 per cent were recorded in the 1950s and 1960s; dramatic increases in the growth rate occurred in the mid-1960s when annual increases of almost 400,000 for the UK affected many aspects of planning policy at the time. Equally dramatic was the down-turn in the very next decade when in the mid-1970s the national population was virtually static – again with planning consequences. A modest upturn took place in the mid-1980s with annual increases exceeding 100,000 (Champion and Townsend (1990*a*)). The 1991 UK population was above the 57 million mark, little more than one million above the 1971 level.

Accompanying these changes have been important shifts in geographical distribution. Again, war-time assumptions have not held. Champion and

Townsend (1990*b*) have suggested that a threefold pattern of change has unfolded over the last quarter of a century, in the context of secular movements over an even longer period of migration trends. One movement is a North-South drift; another is an urban-rural exchange; another concerns suburbanization and decentralization. Rural areas in England and Wales have been much influenced by all three.

The North-South drift had been a feature of the inter-war period when the relative decline of the formerly prosperous coalfield manufacturing areas first became apparent. The population of Greater London increased by two million between the wars, 1.25 million accounted for by inward migration. Post-war planning policy was expected to halt the decline of the regions, stemming population and employment outflow by economic support measures, and for perhaps two decades regional movement of population from economically unfavoured to favoured regions was not a great issue, though a broad continuation of past trends was in evidence. But over the last 20 years or so the North-South divide has been reopened. If the South is defined as the four standard regions of South East, South West, East Anglia and the East Midlands, the net movement of people from North to South rose from 10–20,000 annually in the early 1970s to nearly 70,000 in 1985–6. The South East became particularly attractive to migrants, while other regions and Wales suffered badly; during the 1980s the idea of the South East as an economic engine, with service industries particularly buoyant, was of popular appeal.

The urban-rural shift describes a process whereby major cities lose population particularly from their inner areas, and whole city regions may stop growing. Meanwhile certain rural areas with a long record of depopulation may begin to record population increases. The fastest growing places in England and Wales for the last quarter of a century have been small towns on the fringes or beyond the major metropolitan centres, particularly those benefiting from planned overspill as in New Towns or Expanded Towns. But the smaller country towns, especially those with nearby motorway connections, may also be buoyant, with a new economic base of 'footloose' service and distributive trades. Many rural areas, formerly remote and stagnating, have enjoyed surprising prosperity, gaining population faster than the national average.

Suburbanization is a local, centrifugal movement of population, which has been a century-long feature. The decentralist trend has been particularly strong post-war, and indeed may be seen as a consequence of planning policy given the history of the determination to reduce inner-city densities through redevelopment. Strict green belt policies have helped to contain suburban spread, only to result in a wider territorial search via transport corridors such as the M4 and M11 out of London to meet housing and other demands. At that point suburbanization contributes to the processes of diffuse metropolitan spread at city region scale, and becomes difficult to differentiate from the urban-rural shift described earlier.

As a result of these population movements, the rural settlement geography of England and Wales has undergone marked change since 1945. Settlement policies have had to be progressively adapted to cope with changing circumstances. Rural areas particularly in southern England have experienced the effects of all three migration streams, but rural areas generally have benefited in that over the ten years 1981–1991 many county districts of England and Wales have enjoyed a marked population growth. During this period the population of Cornwall rose by nearly 9 per cent, Somerset by 8 per cent and Devon by nearly 5 per cent; Norfolk and Suffolk recorded rises of around 6 per cent; around the Welsh Borders the counties of Shropshire, Hereford and Worcester and Powys all grew by more than 5 per cent (Abbott, 1992).

The nature of housing demand has clearly changed as seen in chapter 9. The critical factors seem to be associated with a rejection of urban stress and a preference for consumer-friendly environments. The new rural population is therefore a selective one: the retired, the personally mobile (car ownership is critical), the relatively well-to-do and qualified, and the more enterprising families. Land use planning formerly rested on the allocation of land for various purposes; when it came to residential allocations there was little appreciation of the importance as to *whom* the houses would be for. The recognition of the social aspects of town planning changed all that, and rural settlement planning moved from numbers and locations to questions of social preferences and social structures of village communities.

Planners in the 1940s could not have anticipated the extent to which the urbanite would proceed to invade the countryside. Their preferred settlement policy, heavily reliant on criteria for spatial distribution and (as we shall see) design and layout, proved irrelevant to the fact that many village communities would dramatically change in their social composition. A strictly land use ('physical') planning policy shorn of sociological relevance proved a limited exercise. The changing village quickly highlighted the importance of the newcomers: the retired, commuters with very different life styles, people with different needs and those with different calls on local services. Community power structures were subtly affected; farmers and landowners still continued to play major roles, but an earlier fixity of a squire-clergy hierarchy was soon broken. A new social polarization between the affluent and the deprived became apparent.

The classic case study by a geographer-sociologist, R.E. Pahl (1965), reviewed in chapter 9, described the development of commuter villages in the metropolitan fringe in Hertfordshire. It was a study of the spatial and sociological processes inherent in metropolitan expansion, the detail of which Abercrombie had rather glossed over in the decentralization proposals contained in the *Greater London Plan*. The discovery of two communities within the same settlement led to a concern over social polarization and rural deprivation becoming a long running theme in British rural planning (Ambrose, 1974; Shaw, 1979), recently invigorated by difficulties experienced in providing affordable housing.

In addition to the need to change from physical and spatial criteria to social and distributive aspects in rural settlement planning, there has been the requirement to achieve efficiency and meet personal choice in the planning system. To secure the necessary adaptability and flexibility without sacrificing standards or principles is not easy, and rural settlement planning has been obliged to tread warily through a minefield of conflicting views and strongly held opinions. Almost any policy may be welcomed or feared according to the nature of the interest held. Some of the most hard fought battles have concerned proposals to build new villages.

In 1962 Bar Hill was proposed as a new village for Cambridge in a scheme to provide the town with a string of necklace villages in its immediate surrounding hinterland (Moss, 1978). Instead of selecting a declining fenland community for expansion and reinvigoration of its services, a plan for a new settlement was drawn up in 1963 to accommodate 4,000 people. By 1969 when almost 200 houses had been completed, the original developers who had aimed for the upper end of the housing market, sold out and a rather different settlement emerged, but it was still regarded by its opponents as an alien intrusion of urban values into the countryside.

New Ash Green, between Sevenoaks and Gravesend in Kent, was a village scheme for 5–6,000 people by the SPAN corporation, a company specializing in small scale, medium density housing projects in the London area; the architect, Eric Lyons, represented a 'socially-conscious' developer, building in the style of the Modern Movement (Bray, 1981). A planning application was rejected in 1964. On appeal the Enquiry Inspector recommended that it be dismissed, but the Minister, Richard Crosman, on the advice of his officials, gave approval. There was public opposition and there were allegations that the local authorities concerned were hardly supportive, but the first dwellings were on the market by September 1967. After barely a year housing demand had down-turned. The Greater London Council failed to purchase a block of houses in their overspill programme and operations at New Ash Green were suspended in early 1970. The project was taken over by Bovis and completed with rather less innovative housing and a safer marketing image.

Two such disputed and only partially materialized settlements failed to kill off the idea of new villages. Essex County's expansion of Woodham Ferrers kept it alive in the 1970s. In the rather different circumstances of the 1980s proposals for privately funded small new country towns were put forward by Consortium Developments Ltd, with the aim of meeting some of the housing demand in the South East (Consortium Developments, 1988). The scheme was for a number of such settlements for populations of between 12–15,000 people on land up to 1,000 acres allocated for a full range of urban uses. In 1984 a site at Tillingham Hall, Essex, in the Metropolitan Green Belt was selected; in 1985 the planning application was refused and following an appeal, refusal was upheld by the Secretary of State, Nicholas Ridley in 1987. Meanwhile (1986) a second site for a small country town called Foxley Wood at Bramshill in north east Hampshire was selected. This too was refused, and following an appeal, refusal was upheld by the Inspector on grounds of the development's environmental impact, but, in the light of new estimates for housing demand in mid- and north-east Hampshire by the end of the century, Ridley announced that he was 'minded to allow' the proposal. Public consternation had its political effect: in October, Christopher Patten, Ridley's successor, overturned the previous announcement by declaring that he was now 'minded to dismiss the appeal' (Cherry, 1990). A third attempt to build a new settlement, at Stone Bassett in South Oxfordshire, likewise met with no success, and after an interval the Consortium withdrew from the fray.

But it is unlikely that this will be the end of the story. The London and South East Regional Planning Conference (SERPLAN) has been revising upwards the figures of dwellings required in the South East and in consequence the preparation of County Structure Plans, with their all-important allocations for future dwellings, has become a contentious affair – as with Hampshire and Berkshire, at least two counties having disputes with the Department of the Environment over housing targets. A switch in policy was favoured by Michael Heseltine during his incumbency of DoE (1990–92); he supported the idea of a string of mini-new towns on either side of the Thames, in the so-called East Thames Corridor strategy. Meanwhile the DoE's Planning Policy Guidance Note No 3 (1992*a*) on housing suggests that new villages of around 750 units, to larger ones of 4–5,000 dwellings, should be considered as development options in plans. Many new settlement proposals are believed to have been prepared for various parts of the country, and in the circumstances Heseltine's rejection in 1992 of villages around Cambridge, after a long enquiry, was surprising.

VILLAGE DESIGN AND LAYOUT

The Scott Committee Report (1942), anticipating a degree of building in the countryside after the war, put forward clear pointers as to the high design standards that should be achieved in new settlements. (It also stressed high standards for internal design and accommodation provision, but these matters are not the concern of this chapter.) First, sporadic development should cease:

> Though not all country dwellers can live in groups we consider that planning schemes should be so designed as to direct all new settlers into country towns and villages except where they can advance some decisive reason why they should be housed in the open countryside. (para 204)

Next, design:

> . . . the country town or the village, in the English tradition at any rate, is a closely-knit group of buildings. It is well, indeed, that it should be so; that it should in its physically closely-knit character symbolise the socially co-operative basis of the group of people which it houses. We suggest that in the building of new small towns and villages, and in the remodelling of old towns and villages which we hope will be undertaken, this close-knit character should be maintained. The country towns should be as compact as the requirements of healthy living will permit. The villages . . . should be of the 'enclosed' rather than of the 'roadside' type. Thus the village building should be situated round a green or a series of greens in a pattern which is capable of natural expansion. (para 205)

Scott considered that modern villages 'should not attempt to imitate the old but should perform their modern functions in a frankly modern way.' However, clear design principles should be followed, not only in layout but with regard to siting in the landscape and in the question of quality, colour and character of building materials.

These general guidelines, scarcely in contention, formed a useful statement, the elements of which were readily assimilated into the conventional planning wisdom of the day. The rejection of sporadic development fitted easily with the dominant ethic of the protection of the countryside from most forms of building, and County Development Plans incorporated the policy in their general planning proposals. Agricultural dwellings were exempt from planning control, but by and large an open countryside policy has been successfully maintained, with dwellings confined wherever possible to designated village envelopes.

But with design Scott's aspirations have scarcely been met. Government offered no immediate village design guidance at the end of the war, though Thomas Sharp's *The Anatomy of the Village* (1946) became a popular design guide for village planning; a similar text under the title 'The English Village' subsequently appeared as one part of the MHLG's *Design in Town and Village* (1953). Sharp had acted as one of the secretaries to the Scott Committee and his prescriptions for village planning followed the advocacy of the Report.

With some good photographs and well-drawn plans, Sharp reviewed the English village tradition: village siting and plan form, village character and

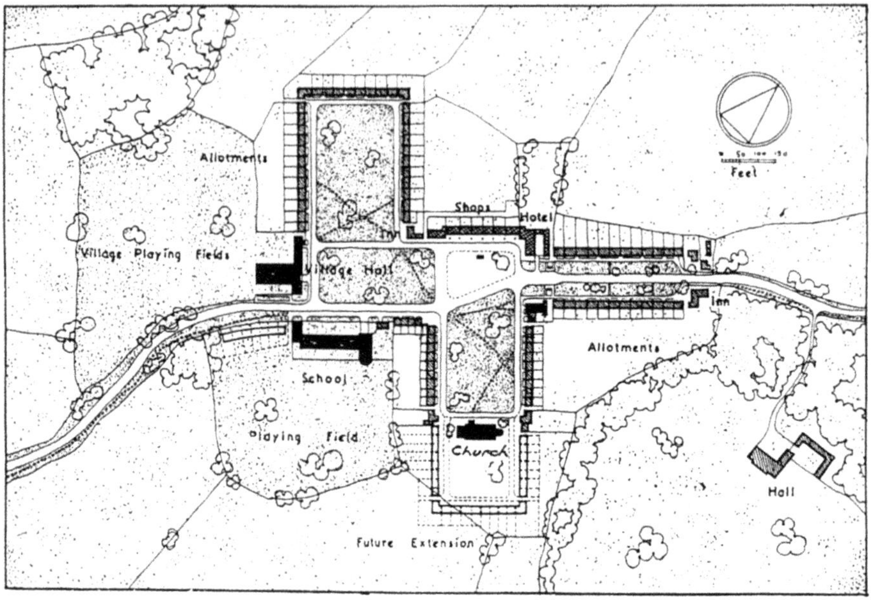

A plan for a new village. A model development by Thomas Sharp in his book *The Anatomy of the Village* (1946), 'showing an inter-play of shapes, and the considered siting of the main buildings'.

its community. He went on to consider future social requirements in villages and set out guidance for future design, emphasizing simplicity of building shapes and spaces between, based for the most part on the rectangle as the most convenient plan form. 'The square, the quadrangle and the close are the most useful plan-shapes for modern conditions of living; and the plans for our new or rebuilt villages, and for village extensions, may very well be based on them, where the topographical conditions are suitable' (p. 63). The principles of scale and enclosure were most likely to be met in this way. On building materials Sharp continued to echo Scott, stressing the importance of colour and texture, to give a village architecture that was robust and colourful.

By the time of the 1953 publication of this essay, the village of Kielder was being built by the Forestry Commission in Northumberland. Sharp had taken the job of planning a considerable number of villages for the Forestry Commission. Of those designed only three were built, and in a very reduced form (Darley, 1975). The stark design matched to some extent the principles he himself enunciated ('the essence of true village character and good village design is simply simplicity', p. 19). But Sharp's passionate, definitive vision for the future village could not match the rising tide of rural building development which the planning system failed to contain beyond a depressing mediocrity. The fact was that there was a lack of design capability in both the private and public sectors, a high threshold of tolerance to aesthetic impoverishment by the general public, and a wariness by planning authorities to turn down building proposals on design grounds. The special issue of the *Architectural Review* entitled 'Outrage', edited by Ian Nairn (1955) attacked the visual products of both urban and rural planning.

Village and small town development failed to heed Sharp's dream of sensitively designed, contained and enclosed villages. There is a strange silence in the literature of the time and it was not really until R.J. Green's *Country Planning* (1971) and A. Thorburn's *Planning Villages* (1971) that the whole question of building in the countryside received fresh attention. Typically in areas of stone building sympathetic building materials would be required by local planning authorities, but it was not until the Essex Design Guide dating from 1973 and its many imitators laid down comprehensive design guidance for buildings, landscaping and spatial composition that any real attempt at coming to terms with the reality of the flood of new development was made. The result has been a post-war rash of urban-style estates, prairie *culs-de-sac* and alien intrusions into past rural forms.

By the end of the 1980s there was a reaction. The context was a response to modernism and post-modernism. Symbolically the architect Leon Krier unveiled his Master Planning Objectives for the Poundbury development at Dorchester on 200 acres of Duchy of Cornwall land, in June 1989: four separate quarters, focusing on a new town centre, and containing squares, avenues and civic buildings will make up the new settlement. The proposal echoed the philosophy expressed in Prince Charles' book *A Vision of Britain: A Personal View of Architecture* (1989). Offering 'ten principles on which we can build' (the place, hierarchy, scale, harmony, enclosure, materials, decoration, art, signs and lights, and community) he asserts that

> The skills, the crafts, the art that went into the architecture of the past are still there – just. But they need to be revived and put to work again, so that we can build cities, towns and villages which better reflect the true aspirations of its people . . . We need design and layout which positively encourage neighbourliness, intimacy and, where possible, a sense of shared belonging to a recognized community. (pp. 155–156)

In different words, this was the passion that had marked the onset of the post-war period, but the professions, the building industry, representative democracy and institutions in the planning system had failed to respond to it. In large part, village and small town extensions have failed to live up to earlier aspirations. Scott and Sharp have been confounded and there has been no successor prophet.

GREEN BELTS

The green belt forms a major feature of the strategy for settlement planning. Once again, there are clear links with developments pre-war, but war-time circumstances (particularly the recommendations of Abercrombie in the *Greater London Plan*) provided an unmistakable fillip. Perhaps surprisingly, however, it was not until the mid-1950s that the decisive impetus to land protection in this way was afforded (Thomas, 1970; Elson, 1986). Unwin's green girdle proposal for the Greater London Regional Planning Committee (see chapter 4) was followed in practice by the counties' green belt estates and the London County Council green belt scheme of 1935. The green belt idea was endorsed by the Scott Committee Report (1942), but its role was markedly conservationist in comparison to what had gone before: 'we conceive the green belt to be a tract of ordinary country, of varying width, around a town and as a tract where the normal occupations of farming or forestry should be continued, so that here, as elsewhere in rural land, the farmer is the normal custodian of the land' (p. 71).

This wider, strategic device found expression in Abercrombie's *Greater London Plan 1944*. The green belt was between 5 and 15 miles wide and included most of the land in the surrounding counties already acquired for green belt purposes, together with other open farm land, to be permanently safeguarded from building development. London acquired its first statutory green belt in the early 1950s as part of the new planning system, the provisions of the Town and Country Planning Act 1947 enabling the land to be protected from development with no cost to local authorities because refusal of planning permission attracted no payment of compensation.

Post-war austerity measures gradually receding, there were now sustained improvements in annual house building rates. Harold Macmillan, Churchill's Minister of Housing and Local Government in the Conservatives' return to power in 1951, presided over targets of more than 300,000 dwellings annually. The peripheries of larger towns and cities were subject to pressures of land availability and in 1955 Duncan Sandys, Macmillan's successor, issued on his own initiative a Departmental Circular (42/55) asking local authorities outside London to consider establishing clearly defined green belts. Their purpose was essentially strategic in a land use sense. No mention was made of amenity considerations or provision for leisure; their objective was to check the further growth of large built-up areas, to prevent neighbouring towns from merging into each other, and to preserve the special character of a town.

The absence of any effective regional planning machinery for resolving inter-authority housing land problems made the Circular popular with local authorities. The procedure was that councils first sought formal approval to define green belts by submitting to the Ministry a sketch plan indicating approximate boundaries. When these plans were found to be soundly based, the authority was permitted to submit a formal proposal as an amendment to its Development Plan. A national green belt map had emerged by the end of the first 'round' in 1963: there were 5,664 square miles of land subject to green belt, dominated by London, but with Birmingham-Coventry, West Riding, the Hampshire Coast and Merseyside-Manchester also with extensive tracts (Elson, 1986).

In the later 1960s the green belt enthusiasm waned somewhat, but revived with strong support from Peter Walker, Secretary of State for the Environment in the Heath Administration, 1970–74. Since then green belts have become more and more permanent features of the land use map. The effects of this extensive, protected reservation has appealed to increasing sections of interest including: the environmental lobby and conservationists generally, the agriculturalists who welcome the 'certainty' of protection from development, householder interests in favoured outer-suburban settings, informal recreationists, local politicians who thrive on the discretionary power presented to them, and planning strategists who would favour a green belt noose to urban spread in order to promote inner city recovery, to limit transport usage and reduce CO_2 emissions. In England the extent of designated land is now more than double that of 1979, having increased from 1.8 million acres to 4.5 million acres, and now representing nearly 14 per cent of the total land area of the country – a larger proportion than that for National Parks or Areas of Outstanding Natural Beauty. Some individual green belts are very large: London obviously, but that for the West Midlands for example at 976 square miles is more extensive than the largest National Park (the Lake District).

Compared with National Parks (where discontent has flared in various *causes célèbres*) and with Areas of Outstanding Natural Beauty (where issues have been low-key), the green belt has provided a continuous setting for conflict, the intensity of which has actually heightened in the past decade. The development industry, typified by the volume house builders, has been ranged against local householder interests and the countryside preservation lobby in which the Council for the Protection of Rural England has proved highly effective. Political support for the green belt has been sustained and consistent refusals of planning permission have succeeded in 'holding the line' against urban spread. A highly distinctive rural fringe geography is the result. The 'stop zone' has not of course halted metropolitan expansion – it has merely hopped over to affect even wider tracts of countryside. In the search for greater flexibility a strategy for planned new communities in the green belt, as at Tillingham Hall in Essex, Foxley Wood in Hampshire and around Cambridge in recent years, failed to gain DoE support, and a tentative peace now prevails. Extreme public sector nervousness prevails over green belt change, especially in the light of the unfavourable response to the rather more liberal green belt Circular in 1984, which had to be withdrawn. Elson (1986) concludes that 'manoeuvring, compromise, and readjustment will continue to typify policy making. Green belt will survive because of its symbolic status; like tax relief on mortgages it is immune to removal' (p. 264).

RURAL PLANNING: ITS KEY ELEMENTS

This chapter has focused on land use planning in the countryside. Key elements have been observed in settlement planning, village design and green belts. But these have to be placed in context, and it will be helpful to sketch a chronology for the last half-century in which all the major features of the rural planning 'system' are introduced.

The Town and Country Planning Act, 1947 bequeathed virtually unfettered development rights to the nation's agricultural and forestry industries. Britain's planning system had a built-in urban orientation, its legislative powers directed to the reconstruction, renewal and the orderly growth of towns and cities. It meant that what stood for rural planning was essentially negative, its objectives being to prevent unwelcome forms of urban development in the countryside. The provisions of the Agriculture Act, 1947 would resuscitate village and small town life and protect the landscape through a farming industry with a secure financial base. The problems of the 1930s (the excesses of urban sprawl, down-at-heel village life and a rural landscape despoiled through inadequate maintenance) were thereby tackled. The Ministry of Agriculture acted as a formidable guardian of the agricultural use of the land against all other uses, while the Ministry of Town and Country Planning would severely limit all forms of building development.

During the 1950s the statutory planning system developed its own momentum; Development Plans were prepared, and day-to-day building operations proposed by a myriad of public and private developers were 'controlled' against their provisions. By and large the anti-urban protectionist line held, though not without its stresses and strains when urban authorities sought land beyond their (often tight) boundaries as *Lebensraum* for their burgeoning housing needs. There were spectacular set-piece confrontations between big cities and surrounding counties, well recorded by Hall (1973) in five case studies: London's western fringes; south Hampshire; Birmingham and the West Midlands and Coventry; Leicester and Leicestershire; and the north west with its two conurbations. The processes of New Town building, town development for overspill schemes and the increasing demand for private housing in the urban fringe were reasonably accommodated however by rigorously controlled, selective land release. Green belt designations strengthened the notion of 'containment' in land use planning. It was known that post-war residential densities in areas of new development were much higher than had been the case between the wars, so overall by the end of the decade the assumptions and policies of the 1940s had not been seriously challenged.

Yet by the early 1960s Best and Coppock (1962) could hazard that 'since 1950 it is probable that well over half a million acres of land have changed their primary use in England and Wales' (p. 232); much of this would represent a net loss to the rural land supply. The 1960s sustained, and in some years exceeded the high annual national building rates already established. Increasing car ownership, the beginnings of a national motorway system and road improvements generally facilitated longer distance commuting. Villages and small towns became targets for housing development. The saving of agricultural land (see chapter 6) was still paramount on the strategic planning agenda – though it would be increasingly asked why, as agricultural productivity soared. Edwards and Wibberley (1971) concluded that Britain would not be short of agricultural land by the end of the century, unpalatable findings which were neglected by MAFF and the NFU for some years.

Meanwhile, other aspects of rural planning had their own momentum for change, particularly in respect of National Parks and countryside recreation and the increasing interest in nature conservation and the environment. There was increasing pressure on the countryside to accept an expanding scale of usage, while at the same time demand for greater protection to withstand encroachment by all kinds of development was more and more vociferous. By the early 1970s there were indications that the lowly status of land use planning with regard to the countryside was at least being reviewed, if not exactly under scrutiny, by practising planners: *Country Planning: The Future of the Rural Regions* was perhaps a turning point (Green, 1971). General books quickly followed, including Cherry (1976) and Davidson and Wibberley (1977).

By now the Town and Country Planning Act, 1968 had modified the 1947 legislation through the introduction of a 'two-tier' system of Structure Plans and Local Plans, but the essential arrangements for planning remained the same. For rural areas an emphasis on environmental management persisted, with powerful rural planning agencies critical in shaping the use and appearance of the countryside: the Ministry of Agriculture with its system of guaranteed prices, grants and advisory services, the Forestry Commission and its activities, and the Nature Conservancy. By contrast, towns and cities were regulated by a system of development planning. Perhaps for 30 years and more the nature and practice of rural planning had these clear characteristics: the key feature was resource planning as opposed to development planning.

And so it remains, even though planning agencies and directives change over the years. Minerals and water provide good examples. The control of mineral working was the subject of a Memorandum first prepared in 1951 as a guide to the planning control of the activity. A revised edition was published in 1960. There have been further changes in the statutory provisions and in government policy, and revised guidance was issued in the DoE Minerals Planning Guidance Notes. Notwithstanding the requirement to provide for the proper winning and working of minerals to contribute to the nation's prosperity, the need 'to preserve the nation's heritage, to improve the quality of the environment, to protect the countryside including the Green Belts and to conserve the best and most versatile agricultural land' remain important objectives (DoE, 1988).

The importance of the water environment has also been stressed in recent years, new standards for river water quality and bathing beaches having been set. Water supply and sewage disposal has been privatized and a National Rivers Authority created from the parts of the former water authorities that dealt with river basin functions.

There is abundant evidence to show that the natural beauty and amenity of the countryside has remained a central concern in rural planning. Sheail (1992*b*) has suggested indeed that the obligation to take account of the amenity, wildlife and out-door recreational interests, imposed on select industries, is 'an insight into half a century of environmental protection'. The 'amenity' clause was first imposed on the hydro-electric power industry in Scotland, legislation in the early 1920s obliging undertakers to have regard for scenic beauty in their operations; the precedent was generalized in 1957 to embrace all the activities of the electricity industry, subsequently extended to other utilities. In a third phase the 'amenity' clauses were central to the Countryside Acts for Scotland (1967), and England and Wales (1968); last, the obligations have been placed on the privatized public utilities.

It is in this context that over the last 15–20 years the resolution of conflict over the use of land has come to represent a critical dimension of countryside planning. It is a consequence of a popularly-supported resource planning system, with a strict control system in force and with a raft of protected land designations in place, facing head-on the increasing demands for development in rural areas. The confining collar of metropolitan green belts has highlighted this feature around the major cities; overall, the ideological legacy of resistance to urban spread has been amplified by the new vigour of anti-development postures in the interests of conservation, so that today a protected countryside (for whatever reason) is a *sine qua non* for various countryside lobbies. Yet the reality of city development is that metropolitan forms of expansion are more

Major developments in the countryside. Sizewell nuclear power station, Suffolk, showing the intrusive effect of this and similar major developments on the rural landscape. (By courtesy of the Countryside Commission)

pronounced than ever. Centrifugal tendencies are spreading the metropolitan influences far and wide: for living space, for recreation, for sand and gravel extraction to fuel a building industry, business parks, retail stores, roads, airports, distribution centres, all serviced by the ubiquitous motor vehicle. As has been the case for over a century, rural England (in particular) is being colonized by urban interests and a protectionist ethic in the planning system is now creaking under the strain.

Pragmatism will prevail. There is increasing recognition of a multipurpose countryside. Agriculture, albeit the dominant user of the land, is shrinking and its labour force will never again be a significant employer in the national economy. New types of commercial activity and industry, widely introduced into a host of villages and country towns, are bringing important changes to the rural economy. The population of the rural areas continues to grow and additional housing is required. Tourism, sport and recreation have their growing demands. In the face of these changes the conclusions of the Scott Report (1942) have long since had to be readdressed and an adaptive planning system called upon to manage the countryside environment. A critical section of the government's policy (DoE, 1992*c*) now reads:

> The guiding principle in the wider countryside is that development should benefit the rural economy and maintain or enhance the environment. The countryside can accommodate many forms of development without detriment, if the location and design of development is handled with sensitivity. New development in rural areas should be sensitively related to existing settlement patterns and to the historic, wildlife and landscape resources of the area. Building in the open countryside, away from existing settlements or from areas allocated for development in development plans, should be strictly controlled. In areas statutorily designated for their landscape, wildlife or historic qualities, policies give greater priority to restraint (para 1.10).

This is a masterpiece of drafting, well within the confines of the flexible, discretionary tradition of British planning. It represents continuity from the past, it acknowledges the need for adjustment to changing circumstances; it recognizes the pillars of a preservationist ethic; it affirms the importance of protected areas, yet it accepts the validity of some forms of development. It enables change to occur, without conceding the force of past policies.

Policies for the countryside, pursued for almost 50 years, fall neatly into the government's environmental strategy, published in 1990 as *This Common Inheritance* (DoE, 1990). The White Paper 'records what we are doing already and what we plan to do to make our air and water cleaner, to preserve the beauties of our countryside and historic buildings' (para 1.39). Specific proposals include a programme of energy efficient measures, attempts to quicken the phase-out of ozone-depleting gases, measures to combat noise pollution, help for historic cathedrals, better use of derelict and vacant land, encouragement to organic farming, grants for woodland management, a new forestry initiative, landscape conservation and a forum to discuss heritage policy. The scope of the White Paper was urban as well as rural, and, such is the canvas, where the professional remit for planning begins and ends is impossible to say. But it does demonstrate the contemporary shift in emphasis for the countryside planning agenda. Meanwhile the statutory processes and the ideological mind set all too often remain firmly placed in the context of the 1940s.

11

RURAL CHANGE AND PLANNING: THREADS IN THE PATTERN

At the beginning of the period covered by this book, the twin concepts of *change* and *planning* which are at its heart would have appeared strange to many commentators. The countryside was seen to have as one of its key characteristics a quality of stability and wholesomeness which clearly differentiated it from the ever-changing, unhealthy cities. It followed that to engage in forms of State regulation for the countryside was not only unnecessary but also meaningless. Eighty years on, neither viewpoint has stood the test of time.

The story which has been unravelled in the preceding chapters has been a complex one, which indeed appears to have become more and more complicated as the century progressed. At various times the countryside has been marginalized from public policy, as for example in the inter-war period and again in the 1960s when urban and industrial problems were judged as of greater import. At other times, for example in the years immediately after the Second World War and again in the 1980s and early 1990s, the countryside has moved much more centre stage as a focus for policy and concern. This increasing complication, coupled with the waxing and waning of policy activity, inevitably make for a complex picture. It is a function of this final chapter to spell out some underlying themes which can help in an understanding of this complexity.

PROTECTIONIST VERSUS DEVELOPMENTAL PRESSURES

In his valedictory essay published at his retirement as Professor of Countryside Planning in the University of London, Gerald Wibberley pointed to the dual strands of protection and development which could be traced throughout the twentieth century in the history of rural planning (Wibberley, 1982). This simple dichotomy proves a useful starting point in this concluding overview of major themes.

To many people, professional and lay alike, the essence of rural planning in a crowded and largely urban land must lie in protecting the 'countryside resource'. The theme is not a new one, and indeed stretches back at least to Tudor times. Concern grew in the nineteenth century such that in 1829 the cartoonist George Cruickshank was railing against 'London going out of town – or the march of bricks and mortar'. The later years of the nineteenth century saw the beginnings of the modern countryside pressure groups (Lowe, 1989). As chapter 8 showed, this particular form of protectionism (an interesting blend of hard politics and neo-populism) has gone from strength to strength, especially in the last few decades.

But protection of the countryside has also been institutionalized and become part of the rationale of the State as rural policies have developed. For much of the period under consideration this rationale focused upon strictly utilitarian needs related to the safeguarding of the land resource underpinning food production. This view was especially important, of course, in the years after the Second World War and was, as chapter 6 particularly emphasized, at the heart of both the agricultural and land use planning policy fields. In more recent years, as food supply seemed more assured and as concern grew at the financial and environmental cost of supporting agriculture, so this protectionist justification has broadened from its narrow utilitarian origins. Thus in the 1980s rural protection took on new meanings linked, for example, to Environmentally Sensitive Areas or Nitrate Protection Zones and government has even gone so far as to recognize the need to safeguard the countryside for its own sake. This modern policy development provides a fascinating throw-back to the wartime years when Thomas Sharp, one of the joint secretaries of the Scott Report, argued strongly for the separation of country from town largely on the grounds of the distinct and unique character of the countryside.

Though it comes as a surprise to many, the developmental strand in rural planning history is scarcely less old. Wibberley (1982) indeed, argued that the creation of the Development Commission in 1909 marked 'the real beginning of conscious thought and work in the broad field of countryside planning.' The Commission is in fact the oldest of the countryside agencies and the progenitor of many rural development initiatives (for example in education, agricultural research and extension work) which are now the accepted brief of other branches of the public service. The need to improve living standards in rural areas was a frequent concern in the interwar period (chapter 4) and was, of course, a key element in the Scott Report. To some extent the developmental emphasis lost some of its impetus in the second half of the century. Rural living standards demonstrably improved as the nation as a whole generally prospered and the rural effort narrowed down to focus upon productivist agriculture. Moreover the creation of separate development agencies for the more rural parts of Scotland (in the

1960s) and Wales (in the 1970s) inevitably forced a redefinition of the developmental need to apply to the upland peripheral areas rather than to the countryside as a whole. In more recent years, and especially after the coming to power of Conservative administrations after 1979, there has been an attempt to ease development restrictions in the countryside as the need to diversify the rural economy has been more and more emphasized, though this has arguably been of limited success and major protectionist policies such as the green belt have, of course, remained firmly in place.

TOWARDS A MULTI-FOCUS COUNTRYSIDE

A second theme which emerges is the change since the beginning of the twentieth century in the perceived role of rural areas. At the outbreak of the Great War and arguably for many decades to come, there was a generally held view that the countryside and farming were synonymous and that the general health of the former was dependent upon the specific health of the latter. All aspects of rural life and landscape seemed naturally to flow from this symbiosis: the stewardship of land, direct and indirect employment, attractive landscapes. The calls for rural revival in the interwar period, the tenor of the Scott Report in 1942 and the thrust of all the legislation in the immediate post-war years were all directed to the same end: the improvement of agriculture as the sole *raison d'être* of the countryside.

In retrospect it is possible to see the tentative emergence of other roles for the countryside. The growth in informal recreation in the period between the Wars (chapter 4) heralded a recreation boom three decades later. The spread of housing along the arterial roads and the growth of industry in small market towns was, in retrospect, a precursor of the counterurban movements of the last quarter of the century. Despite the agricultural depression of the 1920s and 1930s, there were signs even then of the modern agricultural practices, involving chemicals and machinery, which would all but do away with the traditional mixed farming so lauded by the Scott Committee (chapter 5). But at the time these signs were either ignored or were viewed as short-lived exceptions to the dominance of agriculture and the continuity of generations.

No student of rural areas could, at the end of the twentieth century, accept this single-minded view of the countryside. It is not just a question of increasing pressures upon agriculture nor even just the threat of urbanization. Rather is there an acceptance, if sometimes unwilling, that the countryside has many legitimate claimants, not least from the urban population. New housing and new jobs, nature conservation and landscape, farm pluriactivity, recreational access and diversification are all common-place elements in the countryside of the 1990s.

This multi-purpose countryside has been well described as an 'arena' by the Archbishops' Commission on Rural Areas (ACORA, 1990). Not only is this concept useful in that it makes clear the variety of activities which are found in the countryside, but it also encompasses the perception and beliefs of all those who have a stake in the countryside. As the ACORA report says:

> In the 1990s, therefore, England's rural areas may be best understood as an arena – an arena in which different concerns and aspirations, stimulated by social and economic change, meet and must somehow be reconciled. But they are an arena in deeper senses too. For the differences they harbour are not only between interests; they also concern the way the countryside itself is understood.
>
> . . . This reality is much more than a merely theoretical or academic matter. It is reflected at every level of experience – in the multifarious reports on rural issues that appear each year, in the activities of pressure groups, in the policy priorities of governments and local authorities, and in the everyday lives of people (ACORA, 1990, p. 4).

INTEGRATION INTO A WIDER WORLD

A concomitant feature of the farming focus of the countryside at the beginning of the twentieth century was the essential 'localness' of rural life. For most rural residents in the years before the Great War, the economic and social frame within which they lived their lives was bounded by the village and the country town. By the same token, few city dwellers moved beyond their urban bounds to visit the countryside, though the London fringe in particular had its growing band of recreationists. There were links between the two, of course, but they tended to be to the permanent detriment of the rural areas as the countryside depopulated and rural people moved to the town in search of work. As Ravenstein had noted two decades before:

> The inhabitants of the country immediately surrounding a town of rapid growth flock into it; the gaps thus left in the rural population are filled up by migrants from more remote districts, until the attractive force of one of our rapidly growing cities makes its influence felt, step by step, to the most remote corner of the kingdom (Ravenstein, 1885).

Such migration was likely to be permanent as the villager became a town dweller. Rural areas remained isolated from the mainstream of the nation's life and quite literally 'the places left behind'.

In fact the historic isolation and localness of the countryside, though still very real for the vast majority of rural people, had been in flux since the middle of the nineteenth century. From as early as the 1850s the agricultural economy came under threat as ships began to bring cheap grain, and later meat, from the New World. At home, the key was also developments in transportation, particularly railways which in the 1880s and 1890s pushed lines deep into rural England. A symbolic occasion was the opening of the Metropolitan Railway in 1892, an event which allowed the opening-up of much of the countryside of south Buckinghamshire and the growth of small settlements like Amersham and Gerrards Cross. The circumstances were repeated all around London with parallel effects upon the countrysides of Hertfordshire, Essex, Kent and Surrey (Coppock, 1964).

The growth of the railway network was but one of the developments in the technology of communication which have revolutionized rural life in the twentieth century. There is a fascinating paradox to be explored here. Some of the key inventions of the high Victorian period, a period seen as representing the epitome of local 'village England', were to be developed to the full in the twentieth century. The wireless telegraph, the telephone and the motor car were to play a major part in linking town and country thus destroying the localness and bringing the modern world into the countryside.

In fact the process of change has gone further than merely introducing urban influences into the countryside. In practice many rural areas have become the focus and not just the adjunct of new economic and social forces. The fifth Kondratieff cycle (Hall, 1985), the long wave of economic growth associated with the burgeoning industries based upon computers, information systems and biotechnology which has come to fruition in the last two decades, has especially targeted the accessible, and indeed the not-so-accessible, countryside. Indeed it is arguable that the conventional understandings of accessibility (Moseley, 1979) have now to be redefined as electronic communications open up even the most isolated region. Moreover, this accessibility is not just a joining of the village or market town to the national capital. Rather does it now have a European and an international dimension, as rurally-based entrepreneurs link with Brussels and the countryside is invaded by tourists from France, Germany and the Netherlands.

While futuristic new industries in the countryside inevitably capture the public imagination, the apparently more mundane affects the lives of ordinary people. As one of the present authors has argued elsewhere, the lifestyle of the farmworker bears little relationship to that of his grandfather living in the same house 80 years before:

> His wages are still low, relative to his more affluent neighbours, but they are spent on packaged foodstuffs and consumer goods which are the same as those bought by the residents of the big cities. Both he and his family have experienced an educational system which is common to the rest of the nation and which rarely makes reference to their immediate local rural circumstances.

> Each evening the family participates in world events as international news and entertainment from other nations appear on the television screen (Rogers, 1989, p. 103).

WHOSE COUNTRYSIDE?

At the outset of the period reviewed by this book the answer to the question 'Who owns the countryside?' would have been obvious to most people. The aristocracy were apparently still dominant both in terms of the land which they held and the social and political influence which they continued to wield in the village as landlords, benefactors and employers. Moreover rights and privileges associated with land ownership were hardly questioned and the idea that there could be claims upon the countryside from those who had no locus of ownership or birthright would have seemed strange.

In fact the political and economic base of the great landowners had begun to crumble from at least the 1880s. The agricultural depression of the 1880s and 1890s hit hard at the financial returns from farming and, at least while Liberal administrators were in power, demands to break the power of the landowners went higher up the political agenda. A respite in this political pressure from 1886 to 1906, when there were Conservative administrations, simply delayed the inevitable.

The matter of land ownership is perhaps the most obvious, but not the only, indication of the changes in claims on the countryside during the twentieth century. Sales of land, not least to thrifty Scottish farmers, had begun in the 1880s and were a sign that the ownership of the countryside was changing. In 1912 there was the beginning of major sales of land by landlords to tenants and this continued apace after the Great War. To some extent these changes were from a noble landlord to one whose wealth had come from Victorian manufacturing, but many sales were to ordinary tenant farmers. Changes in landownership, and especially the growth of the owner occupier farming his own land, continued into the second half of the century. While there are many parts of rural England where the aristocratic and noble interest can still be found, the pattern of landownership and land tenure, while overwhelmingly private, is now much more varied in terms of the social origins of owners and the existence of the corporate landowner and the professional land agent.

But, as noted above, landownership is but one aspect of the change. Another is the basis of political power in the countryside and in the nation as a whole. At the outset of the period, political power was closely linked with what Cannadine (1990) has called the 'British landed establishment'. In all administrations from 1880 until the outbreak of the Great War, there was never less than 15 per cent of the Cabinet belonging to this establishment, even under Liberal prime ministers. After the Great War, with the exception of Baldwin's 1924 administration (and then only just), the figure never rose above half. More commonly it was under a quarter between the wars and, after 1945, usually no more than three or four Cabinet members out of 20 would be substantial owners of land. At the local level, political power has also changed. Before the Great War, the aristocratic landlord in the 'big house' exercised great influence as magistrate, county councillor, school governor and the like. The changing pattern of landownership noted above inevitably wrought changes on the local political front also. 'Ordinary' farmers, smaller landowners and tradesmen began to take over these roles in the interwar period and, by the 1960s and 1970s (Newby *et al.*, 1978), had become a major force at county, district and parish level. Thereafter the power of the strictly farming landed interest seems to have waned in the last decade or more as population growth has brought in an enthusiastic manager class which is keen to serve the community.

But there is a bigger, if more amorphous, issue about claims to the countryside which has seen revolutionary change over the period. In 1914 the only claims to countryside use and access which were recognized were those which were vested in ownership. Eighty years on that is no longer the case. The evidence is overwhelming that the ordinary urban public feel that they have a stake in 'their' countryside. Especially from the 1920s and 1930s there have been increasing demands for access to the countryside for recreation and leisure, demands which have become to some extent realized in legislation from the late 1940s onwards. The membership of countryside pressure groups has grown enormously in the last third of the century, evidence yet again of a growing commitment by the general public to the countryside. The concerns about the destructive effects of modern agricultural practices are voiced by urban people who are increasingly angry that their taxes have been used to destroy their countryside. In the light of these revolutionary changes, the question posed by Marion Shoard (1987) in her polemic *This Land is our Land*:

> After centuries spent consolidating their power, can we really get our landowners to start yielding access rights, paying over part of their financial gains and allowing the public a say in the destiny of the landscape?

does not appear so impossible.

EXPERTS AND SPECIALISTS

A fifth and final change during the century relates to the broad issue of planning the countryside. The countryside of 1914, owned and ruled by the establishment, focused upon farmers and farming and linked often only tenuously with the wider world, was not seen as an arena for State intervention and economic management. The countryside of the 1990s is the focus of attention for planners, managers and experts of all sorts. From agricultural policy to nature conservation, from recreational use to community development, the countryside experts exist in their legions. Three examples will suffice to illustrate this trend.

The application of the sciences, particularly chemistry, to the practice of agriculture had made significant strides in Victorian times (Sykes, 1981), but it was really the agricultural depression of the 1880s and 1890s which strengthened the argument that farming could benefit from scientific research. At broadly the same time, as Richards (1994) has shown in his centenary history of Wye College, there were parallel moves to formalize systems of agricultural education. A third impetus was the creation of the Development Commission in 1909 with a specific brief which clearly encompassed agricultural modernization. Thus by the outbreak of the Great War there was a

> . . . new atmosphere of optimism and official belief in science . . . [and in] the principle that only on the basis of State support for education and research could agriculture hope to regain something of the health and economic importance employed fifty years before (Richards, 1994).

The growth of agricultural education and science continued during the interwar period, if somewhat desultorily. The real flowering, however, occurred after 1945. Scientific research expanded both in the universities and in State research establishments. The numbers of scientists, and economists and management specialists too, produced by the universities and by county-based agricultural colleges grew also. And on the government side the State-funded extension services carried the new ideas direct to the farms.

Agricultural science probably illustrates the growth in the role of the expert best, but particularly in the last three decades the agronomists and farm management experts have been joined by other experts. Ecologists and land managers, foresters and community workers have all added to a veritable army of expertise keen to advise on the future of rural areas.

A second illustration of the growth of formalized State involvement can be seen in the growth of the 'quango', the officially appointed group of experts, often recruited from the 'great and the good' and asked to deliberate and advise on rural problems. The idea is an old one, of course; what is different now is the proliferation of these bodies.

While rural panels had existed before (the Board of Agriculture, for example, had been created in 1793), the original 'rural quango' with concern for the whole countryside was the Development Commission, founded in 1909. In its turn it spawned agencies and expert groups, over a wide area of rural concerns, as for example in the creation of the forerunner of the later Agricultural and Food Research Council. Over the century more and more such committees have been created. Some exist in a permanent way, such as the Forestry Commission (founded in 1919), the Countryside Commission (originating as the National Parks Commission in 1949) and what is now English Nature (created as the Nature Conservancy in 1949) together with the equivalent body in the Principality, the Countryside Committee for Wales. Others operate for a limited period only, focusing on a particular problem of the moment. Examples of these are the Northfield Committee on land ownership (1979) and, of course, the Scott Committee (1942) which was so influential in planning the countryside after 1945.

Thirdly, it is appropriate to recognize the growth in the role of the land use planner as far as the countryside is concerned. While planning tended for the first half of the century to concentrate upon the urban environment, there emerged conventional wisdom that the countryside needed special protection because of its intrinsic qualities, as in the case of the Surrey Hills or the South Downs (Sheail, 1981). The real growth in professional planning activity naturally occurred after 1947 and the requirement for all local authorities to produce statutory development plans for all their areas and not just for the urban centres.

It has to be said, however, that relatively few chartered planners expressed much interest in rural matters since urban problems were seen as more pressing. Especially in the public sector, countryside planning was often viewed as at best a tangential interest and at worst a professional backwater. While there were exceptions to this view, for example Ray Green in Norfolk, the prejudice continued until the 1970s. Then, with the growth of interest in conservation and wider environmental issues, with the reduction in the domination of agriculture, the growth of recreational pressures and the emergence of social issues such as rural housing brought on by the counterurbanization trend, the tide changed. The nature of the statutory planning system in England and Wales makes it inevitable that countryside matters are now firmly within the remit of professional activity.

* * * * *

At the midway point of the period covered by this book, C.S. Orwin wrote a small book on *Problems of the Countryside* (1945), intended as a contribution to post-war reconstruction. He began by imagining a latter-day Rip Van Winkle, going to sleep in an English village in about 1880 and awaking some 60 years later to observe the changes which had taken place. The picture which Orwin painted of the countryside of the early 1940s records areas of stagnation and moribund life in the villages coupled with a new liveliness to be found in the country towns.

Orwin saw hope in these contrasts and believed that they signalled the gradual extension of urban standards of living to the deeper countryside. But he equally believed that this would not happen automatically. As he said (p. 14) 'this is a time of planning', and he had no doubts as to the necessity of state involvement in improving the rural lot. At the end of his book, he again imagines Rip Van Winkle returning, this time a generation further into the century. The picture he sees is of a benevolent and successful countryside, efficient yet still attractive, socially vibrant yet ecologically healthy. To the present day reader this picture may appear as utopian and even a little sentimental, used as he is to doom-laden prophecies predicting imminent disaster. Yet, for all the current concerns expressed about the health of the countryside, much of Orwin's imaginings ring true if matched with the countryside of the 1990s. At a time when it is commonplace, indeed almost expected, that rural commentators should stress the problems and the conflicts, it is perhaps salutary to end this account of rural planning and change as Orwin ended his.

> [there was] an impression of a virile, well-knit society, as though there had been a blood transfusion into the old body corporate, which had caused it to expand and to develop, both physically and mentally. There was a vigour and activity about the place which it had never suggested as he remembered it, and he found it good (Orwin, 1945).

BIBLIOGRAPHY

Aalen, F.H.A. (1989) Lord Meath, city improvement and social imperialism. *Planning Perspectives*, **4** (2), pp. 127–152.

Abbott, S. (1992) Rural retreats. *Geographical Magazine*, **LXIV** (1), p. 5.

Abercrombie, P. (1945) *Greater London Plan 1944*. London: HMSO.

Abelson, E. (1988) *A Mirror of England: An Anthology of the Writings of H.J. Massingham (1888–1952)*. Bideford: Green Books.

Acland, A.H.D. (1913) Introduction. Land Enquiry Committee Report *The Land*. London: Hodder and Stoughton.

ACORA (1990) *Faith in the Countryside*. Report of the Archbishops' Commission on Rural Areas. Worthing: Churchman Publishing.

Adams, Thompson and Fry (1930) *North East Kent Regional Planning Scheme*. London: NE Kent Joint Town Planning Committee.

Addison, P. (1987) The road from 1945, in Hennessy, P. and Selden, A. (eds.) *British Government from Attlee to Thatcher*. Oxford: Basil Blackwell.

Allingham, M. (1941) *The Oaken Heart*. London: Michael Joseph.

Ambrose, P. (1974) *The Quiet Revolution: Social Change in a Sussex Village 1871–1981*. London: Chatto and Windus.

Armstrong, W.A. (1988) *Farmworkers: A Social and Economic History, 1770–1980*. London: Batsford.

Armstrong, W.A. (1989) The most despised craftsmen: farmworkers in the twentieth century, in Mingay, G.E. (ed.) *The Vanishing Countryman*. London: Routledge, pp. 115–132.

Arvill, R. (1967) *Man and Environment: Crisis and the Strategy of Choice*. Harmondsworth: Penguin.

Association of County Councils (1979) *Rural Deprivation*. London: ACC.

Balcombe, R.J. (1980) The Rural Transport Experiments: Summary and Conclusion. Proceedings of a Symposium held at the Transport and Road Research Laboratory, Crowthorne, pp. 94–103.

Banister, D. and Norton, F. (1988) The role of the voluntary sector in the provision of rural services – the case of transport. *Journal of Rural Studies*, **4**, pp. 57–71.

Barlow Report (1940) *Report of the Royal Commission on the Distribution of the Industrial Population*. Cmd 6153. London: HMSO.

Barnett, L. (1985) *British Food Policy during the First World War*. London: George Allen and Unwin.

Baron, S. (ed.) (1944) *Country Towns in the Future England*. Report of the Country Towns Conference of the Town and Country Planning Association. London: Faber and Faber.

Barr, C., Howard, D., Bunce, B., Gillespie, M. and Hallam, C. (1991) *Changes in Hedgerows in Britain between 1984 and 1990*. Grange-over-Sands: Institute of Terrestrial Ecology.

Bateman, J. (1883) *The Great Landowners of Great Britain and Ireland*. London: Harrison.
Beeching Report (1963) *The Reshaping of British Railways*. London: HMSO.
Beevers, R. (1988) *The Garden City Utopia: A Critical Biography of Ebenezer Howard*. London: Macmillan.
Bell, P. and Cloke, P. (1990) *Deregulation and Transport: Market Forces in the Modern World*. London: David Fulton.
Best, R.H. (1959) *The Major Land Uses of Great Britain*. Ashford: Wye College, University of London.
Best, R.H. (1981) *Land Use and Living Space*. London: Methuen.
Best, R.H. and Coppock, J.T. (1962) *The Changing Use of Land in Britain*. London: Faber and Faber.
Best, R.H. and Rogers, A.W. (1973) *The Urban Countryside: The Land Use Structure of Small Towns and Villages in England and Wales*. London: Faber and Faber.
Bielckus, C.L., Rogers, A.W. and Wibberley, G.P. (1972) *Second Homes in England and Wales*. Ashford: Wye College, University of London.
Blacksell, M. and Gilg, A. (1981) *The Countryside: Planning and Change*. London: George Allen and Unwin.
Blatchford, R. (1894) *Merrie England*. London: Clarion.
Blunden, J. and Curry, N. (1988) *A Future for Our Countryside?* Oxford: Blackwell.
Blunden, J. and Curry, N. (1990) *A People's Charter? Forty years of the National Parks and Access to the Countryside Act, 1949*. London: HMSO.
Blythe, R. (1969) *Akenfield*. Harmondsworth: Penguin.
Body, R. (1982) *Agriculture: The Triumph and the Shame*. London: Temple Smith.
Body, R. (1984) *Farming in the Clouds*. London: Temple Smith.
Boucher, S., Flynn, A. and Lowe, P. (1991) The politics of rural enterprise: a British case study, in Whatmore, S., Lowe, P. and Marsden, T. (eds.) *Rural Enterprise: Shifting Perspectives on Small-Scale Production*. London: David Fulton, pp. 120–140.
Bourne, G. (1912) *Change in the Village*. London: Duckworth.
Bowers, J.K. and Cheshire, P. (1983) *Agriculture, the Countryside and Land Use: An Economic Critique*. London: Methuen.
Bowler, I. and Lewis, G. (1991) Community involvement in rural development: the example of the Rural Development Commission, in Champion, T. and Watkins, C. (eds.) *People in the Countryside: Studies of Social Change in Rural Britain*. London: Paul Chapman, pp. 160–177.
Bracey, H.E. (1952) *Social Provision in Rural Wiltshire*. London: Methuen.
Bracey, H.E. (1956) A rural component of centrality applied to six southern counties in the United Kingdom. *Economic Geography*, **32**, pp. 38–50.
Bracey, H.E. (1959) *English Rural Life: Village Activities, Organization and Institutions*. London: Routledge and Kegan Paul.
Bracey, H.E. (1972) *People and the Countryside*. London: Routledge and Kegan Paul.
Brasnett, M. (1969) *Voluntary Social Action*. London: Bedford Square Press.
Bray, C. (1981) New Villages: Case Studies, No 1 New Ash Green. Working Paper 51, Department of Town Planning, Oxford Polytechnic, Oxford.
Britton, D.K. (ed.) (1990) *Agriculture in Britain: Changing Pressures and Policies*. Wallingford: CAB International.

Brown, J. (1987) *Agriculture in England: A Survey of Farming 1870–1947*. Manchester: Manchester University Press.
Brown, J. and Ward, S. (1990) *The Village Shop*. Moffat: Rural Development Commission/Cameron and Hollis.
Buchanan, C.D. (1958) *Mixed Blessing: The Motor in Britain*. London: Leonard Hill.
Buchanan, S. (1982) Power and planning in rural areas: preparation of the Suffolk County Structure Plan, in Moseley, M.J. (ed.) *Power, Planning and People in Rural East Anglia*. Norwich: University of East Anglia, pp. 1–20.
Buller, H. and Lowe, P. (1982) Politics and class in rural preservation: a study of the Suffolk Preservation Society, in Moseley, M.J. (ed.) *Power, Planning and People in Rural East Anglia*. Norwich: University of East Anglia, pp. 21–41.
Bunce, M. (1994) *The Countryside Ideal: Anglo-American Images of Landscape*. London: Routledge.
Burrell, A., Hill, B. and Medland, J. (1990) *Agrifacts – A Handbook of UK and EEC Agricultural and Food Statistics*. Hemel Hempstead: Harvester Wheatsheaf.
Calder, A. (1991) *The Myth of the Blitz*. London: Cape.
Cannadine, D. (1990) *The Decline and Fall of the British Aristocracy*. New Haven: Yale University Press.
Chalklin, C.W. (1989) The decline of the country craftsmen and tradesmen, in Mingay, G.E. (ed.) *The Vanishing Craftsman*. London: Routledge, pp. 133–141.
Chambers, J.D. and Mingay, G.E. (1966) *The Agricultural Revolution 1750–1880*. London: Batsford.
Champion, A.G. (ed.) (1989) *Counterurbanization: The Changing Pace and Nature of Population Deconcentration*. London: Edward Arnold.
Champion, A.G. and Townsend, A.R. (1990*a*) Demographic forces and the reshaping of rural England in the late twentieth century. Appendix D of *Faith in the Countryside*, the Report of the Archbishops' Commission on Rural Areas. Worthing: Churchman Publishing, pp. 349–359.
Champion, A.G. and Townsend, A.R. (1990*b*) *Contemporary Britain: A Geographical Perspective*. London: Edward Arnold.
Charles, Prince of Wales (1989) *A Vision of Britain: A Personal View of Architecture*. London: Doubleday.
Cherry, G.E. (1974) *The Evolution of British Town Planning: A History of Town Planning in the United Kingdom*. Leighton Buzzard: Leonard Hill.
Cherry, G.E. (1975) *Environmental Planning 1939–1969: Volume II. National Parks and Recreation in the Countryside*. London: HMSO.
Cherry, G.E. (ed.) (1976) *Rural Planning Problems*. London: Leonard Hill.
Cherry, G.E. (ed.) (1981) *Pioneers in British Planning*. London: Architectural Press.
Cherry, G.E. (1982) *The Politics of Town Planning*. Harlow: Longman.
Cherry, G.E. (1983) Thomas Sharp: The Man who dared to be Different. Sharp Memorial Lecture, University of Newcastle-upon-Tyne.
Cherry, G.E. (1988) *Cities and Plans: The Shaping of Urban Britain in the Nineteenth and Twentieth Centuries*. London: Edward Arnold.
Cherry, G.E. (1990) Town and country planning, in Catterall, P. (ed.) *Contemporary Britain: An Annual Review 1990*. Oxford: Blackwell, pp. 405–411.
Cherry, G.E. (1993) Milestones and signposts in twentieth century planning. *Town and Regional Planning* (Journal of the South African Institute of Town and Regional Planners), **34**, pp. 3–9.

Cherry, G.E. (1994) *Birmingham: A Study in Geography, History and Planning*. Chichester: Wiley.

Cherry, G.E., Jordan, H. and Kafkoula, K. (1993) Gardens, civic art and town planning: the work of Thomas H. Mawson. *Planning Perspectives*, **8** (3), pp. 307–332.

Cherry, G.E. and Sheail, J. (forthcoming) *The Urban Impact on the Countryside, 1850–1914*. Agrarian History, Vol VII. Cambridge: Cambridge University Press.

Christaller, W. (1933) *Die Zentralen Orte in Suddeutschland*. Jena: Fischer.

Clark, D.M. (1990) Affordable Rural Housing – A National Survey of Need and Supply. ACRE for the Rural Development Commission, Cirencester.

Clark, D.M. (1992) Rural Social Housing – Supply and Trends. ACRE for the Rural Development Commission, Cirencester.

Clark, D. and Woollett, S. (1990) *English Village Services in the Eighties*. London: Rural Development Commission.

Clark, G. (1991) People working in farming – the changing nature of farmwork, in Champion, T. and Watkins, C. (eds.) *People in the Countryside: Studies of Social Change in Rural Britain*. London: Paul Chapman, pp. 67–83.

Clarkson, S. (1980) *Jobs in the Countryside: Some Aspects of the Work of the Rural Industries Bureau and the Council for Small Industries in Rural Areas 1910–1979*. Ashford: Wye College, University of London.

Cloke, P. (1979) *Key Settlements in Rural Areas*. London: Methuen.

Cloke, P. (1983) *An Introduction to Rural Settlement Planning*. London: Methuen.

Cloke, P. (1992) *Policy and Change in Thatcher's Britain*. Oxford: Pergamon.

Cloke, P. and Little, J. (1990) *The Rural State? Limits to Planning in Rural Society*. Oxford: Clarendon Press.

Cloke, P. and McLaughlin, B. (1989) Politics of the alternative land use and rural economy (ALURE) proposals in the UK: crossroads or blind alley? *Land Use Policy*, **6** (3), pp. 235–248.

Cloke, P., Milbourne, P. and Thomas, C. (1994) *Lifestyles in Rural England*. Rural Research Series 18. London: Rural Development Commission.

Cloke, P. and Thrift, N. (1987) Intra-class conflict in rural areas. *Journal of Rural Studies*, **3**, pp. 321–333.

Clout, H. (1993) European Experience of Rural Development. Strategy Review Topic Paper 5. Rural Development Commission, London.

Coleman, A. (1977) Land use planning: success or failure? *Architects' Journal*, **165**, pp. 93–134.

Collingridge, J.H. (1987) Parish Government in Rural England: A Study of Present Day Trends and Practices. Unpublished PhD thesis, University of Birmingham.

Commission of the European Communities (CEC) (1988) *The Future of Rural Society*. Brussels: CEC.

Commission of the European Communities (CEC) (1989) *Guide to the Reform of the Community's Structural Funds*. Brussels: CEC.

Connell, J. (1974) The metropolitan village: spatial and social pressures in discontinuous suburbs, in Johnson, J.H. (ed.) *Suburban Growth: Geographical Pressures at the Edge of the City*. Chichester: Wiley, pp. 77–100.

Connell, J. (1978) *The End of Tradition: Country Life in Central Surrey*. London: Routledge and Kegan Paul.

Consortium Developments (1988) *Report on the Plan for Small New Country Towns*. London: Consortium Developments.

Coppock, J.T. (1964) Dormitory settlements around London, in Coppock, J.T. and Prince, H. (eds.) *Greater London*. London: Faber, pp. 265–291.

Coppock, J.T. (1971) *An Agricultural Geography of Great Britain*. London: G. Bell.

Coppock, J.T. (1973) The changing face of England: 1850–1900, in Darby, H.C. (ed.) *A New Historical Geography of England*. Cambridge: Cambridge University Press, pp. 595–673.

Cornish, V. (1937) *The Preservation of Our Scenery: Essays and Addresses*. Cambridge: Cambridge University Press.

Council for the Protection of Rural England (1972) *The Lost Land 1945–1990*. London: CPRE.

Countryside Commission (1968) *The Coasts of England and Wales: Measurement of Use, Protection and Development*. London: HMSO.

Countryside Commission (1987) *Forestry in the Countryside*. Cheltenham: Countryside Commission.

Countryside Commission (1991) *Landscape Change in the National Parks*. Levenshulme, Manchester: Countryside Commission.

Cox, A. (1979) *Brickmaking: A History and Gazetteer*. London: Bedfordshire County Council/Royal Commission on Historical Monuments.

Cox, G. and Lowe, P. (1983) A battle not the war: the politics of the Wildlife and Countryside Act. *Countryside Planning Yearbook*, Vol. 4, pp. 48–76, Norwich: GeoBooks.

Cox, G. Lowe, P. and Winter, M. (1989*a*) *Agriculture: People and Policies*. London: Allen and Unwin.

Cox, G., Lowe, P. and Winter, M. (1989*b*) The farm crisis in Britain, in Goodman, D. and Redclift, M. (eds.) *The International Farm Crisis*. London: Macmillan.

Crawford, A. (1985) *C.R. Ashbee: Architect, Designer and Romantic Socialist*. New Haven and London: Yale University Press.

Creasey, J.S. and Ward, S.B. (1984) *The Countryside between the Wars*. London: Batsford.

Crichton, R. (1964) *Commuters' Village: A Study of Community and Commuting in a Berkshire Village*. Dawlish: David and Charles.

Crouch, D. and Ward, C. (1988) *The Allotment: Its Landscape and Culture*. London: Faber and Faber.

Cullingworth, J.B. (1975) *Environmental Planning 1939–1969: Volume I. Reconstruction and Land Use Planning 1939–1947*. London: HSMO.

Cumbria and Lake District Joint Structure Plan (1993) *Report of Examination in Public*. Carlisle: Cumbria County Council.

Curry, N. (1993) Countryside Planning: Look Back in Anguish. Inaugural Lecture, Cheltenham and Gloucester College of Higher Education, Cheltenham.

Curry, N. (1994) *Countryside Recreation, Access and Land Use Planning*. London: Spon.

Dakers, C. (1987) *The Countryside at War*. London: Constable.

Darby, H.C. (ed) (1973) *A New Historical Geography of England*. Cambridge: Cambridge University Press.

Darley, G. (1975) *Villages of Vision*. London: Architectural Press.

Davidson, J. and Wibberley, G. (1977) *Planning and the Rural Environment*. Oxford: Pergamon Press.

Day, G., Rees, G. and Murdoch, J. (1989) Social change, rural localities and the state: the restructuring of rural Wales. *Journal of Rural Studies*, **5**, pp. 227–244.

Denman, D.R. (1980) *Land in a Free Society*. London: Centre for Policy Studies.

Department of the Environment (1980) *Development Control – Policy and Practice*. Circular 22/80. London: DoE.

Department of the Environment, Welsh Office (1988) *Minerals Planning Guidance*. MPGI General Considerations and the Development Plan System. London: HMSO.

Department of the Environment (1990) *This Common Inheritance: Britain's Environmental Strategy*. Cmd 1200. London: HMSO.

Department of the Environment (1991) *Planning and Affordable Housing*. Circular 7/91. London: HMSO.

Department of the Environment (1992*a*) *Housing*. Planning Policy Guidance 3 (revised). London: HMSO.

Department of the Environment (1992*b*) *Coastal Planning*. Planning Policy Guidance 20. London: HMSO.

Department of the Environment (1992*c*) *The Countryside and the Rural Economy*. Planning Policy Guidance 7. London: HMSO.

Derounian, J.G. (1993) *Another Country: Real Life beyond Rose Cottage*. London: NCVO Publications.

Dewey, P.E. (1989) *British Agriculture in the First World War*. London: Routledge.

Dickinson, R.E. (1932) The distribution and functions of the smaller urban settlement of East Anglia. *Geography*, **17**, pp. 19–31.

Douglas, R. (1976) *Land, People and Politics: A History of the Land Question in the United Kingdom 1878–1952*. London: Alison and Busby.

Dower, J. (1945) *National Parks in England and Wales*. Cmd 6628. London: HMSO.

Downing, P. and Dower, M. (1972) *Second Homes in England and Wales: An Appraisal Prepared for the Countryside Commission*. Cheltenham: Countryside Commission.

Dudgeon, P. (ed.) (1989) *Village Voices: A Portrait of Change in England's Green and Pleasant Land, 1915–1990*. London: Sidgwick and Jackson.

Dunn, M.C. (1973) Patterns of population movement in Herefordshire: implications for rural planning. Paper presented to IBG Annual Conference, Birmingham; quoted in Cherry, G.E. (ed.) *Rural Planning Problems*. London: Leonard Hill.

Edwards, A. and Wibberley, G.P. (1971) *An Agricultural Land Budget for Britain 1965–2000*. Ashford: Wye College, University of London.

Elson, M.J. (1986) *Green Belts: Conflict Mediation in the Urban Fringe*. London: Heinemann.

Errington, A. (1988) Disguised unemployment in British agriculture. *Journal of Rural Studies*, **4**, pp. 1–7.

Errington, A. (1990) Investigating rural employment in England. *Journal of Rural Studies*, **6** (1), pp. 67–84.

Fairbrother, N. (1970) *New Lives, New Landscapes*. London: Architectural Press (Harmondsworth: Penguin, 1972).

Fielding, A.J. (1992) Migration and social mobility: South East England as an escalator region. *Regional Studies*, **26** (1), pp. 1–15.

Flynn, A. (1986) Agricultural policy and party politics in post-war Britain, in Cox, G., Lowe, P. and Winter, M. (eds.) *Agriculture: People and Policies*. London: Allen and Unwin, pp. 216–236.

Fothergill, S. and Gudgin, G. (1982) *Unequal Growth: Urban and Regional Employment Change in the UK*. London: Heinemann.

Fowler, P.J. (1987) The contemporary past, in Wagstaff, J.M. (ed.) *Landscape and Culture: Geographical and Archaeological Perspectives*. Oxford: Blackwell, pp. 173–191.

Francis, D. and Henderson, P. (1992) *Working with Rural Communities*. London: Macmillan.
Frankenberg, R. (1966) *Communities in Britain: Social Life in Town and Country*. Hardmondsworth: Penguin.
Garbett-Edwards, D.P. (1972) The establishment of new industries (with particular reference to recent experience in Mid-Wales), in Ashton, J. and Long, W.H. (eds.) *The Remoter Rural Areas of Britain*. Edinburgh: Oliver and Boyd, pp. 50–73.
Gardner, T.W. (1979) Agricultural Policy: Formative Influences in Britain. Bulletin No. 169. Department of Agricultural Economics, University of Manchester.
Gaskell, P.T. and Tanner, M.F. (1991) Agricultural change and Environmentally Sensitive Areas. *Geoforum*, **22** (1), pp. 81–90.
Gasson, R.M. (1966*a*) The challenge to British farming, 1960–1970. *Westminster Bank Review*, pp. 32–41.
Gasson, R.M. (1966*b*) *The Influence of Urbanization on Farm Ownership and Practice*. Ashford: Wye College, University of London.
Gasson, R.M. (1988) *The Economics of Part-Time Farming*. London: Longman.
Gasson, R.M. (1992) Farmers' wives: their contribution to the farm business. *Journal of Agricultural Economics*, **43**, pp. 74–87.
Gasson, R.M. and Hill, P. (1990) *An Economic Evaluation of the Farm Woodland Scheme*. Ashford: Wye College, University of London.
Gasson, R.M., Shaw, A. and Winter, M. (1992) Characteristics of Farm Household Pluriactivity in East and Mid Devon. Occasional Paper No 19, Centre for Rural Studies, Royal Agricultural College, Cirencester.
Gasson, R. and Errington, A. (1993) *Farm Family Business*. Wallingford: CAB International.
George, H. (1880) *Progress and Poverty*. London: Kegan Paul.
Gibbons, S. (1932) *Cold Comfort Farm*. London: Longman.
Gittins Report (1967) *Primary Education in Wales*. Central Advisory Council for Education. London: HMSO.
Glyptis, S. (1991) *Countryside Recreation*. London: Longman.
Glyptis, S. (ed.) (1993*a*) *Leisure and the Environment – Essays in Honour of Professor J.A. Patmore*. London: Belhaven.
Glyptis, S. (1993*b*) *Planning and Rural Recreation in Britain*. Aldershot: Avebury.
Goldsmith, E. *et al.* (1972) Blueprint for survival. *The Ecologist*, **2** (1), pp. 1–43.
Grant, I. and Maddren, N. (1975) *The Countryside at War*. London: Jupiter.
Grassie, J. (1983) *Highland Experiment: The Story of the Highlands and Islands Development Board*. Aberdeen: Aberdeen University Press.
Green, B.H. (1985) *Countryside Conservation: The Protection and Management of Amenity Ecosystems*. 2nd edn. London: Allen and Unwin.
Green, O. (1987) Introduction. *Metroland* (1932 edition). Harpenden: Oldcastle Books.
Green, R.J. (1971) *Country Planning: The Future of the Rural Regions*. Manchester: Manchester University Press.
Gregory, R. (1971) *The Price of Amenity*. London: Macmillan.
Grigg, D. (1987) Farm size in England and Wales, from early Victorian times to the present. *Agricultural History Review*, **35** (2), pp. 179–189.
Gudgin, G. (1990) Beyond farming: economic change in rural areas of the United Kingdom. Appendix F of ACORA *Faith in the Countryside*. Report of the

Archbishops' Commission on Rural Areas. Worthing: Churchman Publishing, pp. 369–380.

Haggard, H.R. (1906) *Rural England*, 2 vols. London: Longmans, Green.

Hall, A.D. (1913) *A Pilgrimage of British Farming*. London: Murray.

Hall, A.D. (1941) *Reconstruction and the Land: An Approach to Farming in the National Interest*. London: Macmillan.

Hall, P. (1973) *The Containment of Urban England*, 2 vols. London: George Allen and Unwin.

Hall, P. (1985) The geography of the fifth Kondratieff cycle, in Hall, P. and Markensen, A. (eds.) *Silicon Landscapes*. London: Allen and Unwin, pp. 1–19.

Hammond, J.L. and B. (1911) *The Village Labourer*. London: Longman.

Hammond, R.J. (1951) *Food*. History of the Second World War. London: HMSO.

Hardy, D. (1979) *Alternative Communities in Nineteenth Century England*. London: Longman.

Hardy, D. (1991) *From Garden Cities to New Towns: Campaigning for Town and Country Planning 1899–1946*. London: E. and F.N. Spon.

Hardy, D. and Ward, C. (1984) *Arcadia for All: The Legacy of a Makeshift Landscape*. London: Mansell.

Harley, J.B. (1975) *Ordnance Survey Maps: A Descriptive Manual*. Southampton: Ordnance Survey.

Harper, S. (1989) The British rural community: an overview of perspectives. *Journal of Rural Studies*, **5**, pp. 161–184.

Harper, S. (ed.) (1993) *The Greening of Rural Policy: International Perspectives*. London: Belhaven.

Harrison, M. (1981) Housing and town planning in Manchester before 1914, in Sutcliffe, A.R. (ed.) *British Town Planning: The Formative Years*. Leicester: Leicester University Press, pp. 106–153.

Havinden, M. (1981) The model village, in Mingay, G.E. (ed.) *The Victorian Countryside*. London: Routledge and Kegan Paul, pp. 414–427.

Hayek, F.A. (1944) *The Road to Serfdom*. Chicago: University of Chicago Press.

Henderson, P. and Francis, D. (eds.) (1993) *Rural Action: A Collection of Community Work Case Studies*. London: Pluto Press.

Hennessy, P. and Seldon, A. (eds.) (1987) *Ruling Performance: British Government from Attlee to Thatcher*. Oxford: Basil Blackwell.

Hill, N.W.B. (1991) A European Community view of changes in rural society. Paper given at the Eleventh Agricola Conference, Wye College, University of London, Ashford.

Hill, B. and Young, N. (1991) Support policies for rural areas in England and Wales. *Journal of Rural Studies*, **7**, pp. 191–206.

Hirsch, F. (1977) *The Social Limits to Growth*. London: Routledge and Kegan Paul.

H.M. Treasury (1976) *Rural Depopulation – Report by an Interdepartmental Study Group*. London: HMSO.

Hobhouse Report (1947) *Report of the National Parks Committee (England and Wales)*. Cmd 7121. London: HMSO.

Hodge, I. and Monk, S. (1991) *In Search of a Rural Economy: Patterns and Differentiation in Non-Metropolitan England*. Cambridge: Department of Land Economy, University of Cambridge.

Horn, P. (1976) *Labouring Life in the Victorian Countryside*. Dublin: Gill and Macmillan (Gloucester: Alan Sutton, 1987).

Horn, P. (1984) *Rural Life in England in the First World War*. Dublin: Gill and Macmillan.

Hoskins, W.G. (1955) *The Making of the English Landscape*. London: Hodder and Stoughton (Harmondsworth: Penguin, 1970).

House, J.W. (1976) The geographer and policy making in marginal rural areas: the North Pennines Rural Development Board, in Coppock, J.T. and Sewell, W.R.D. (eds.) *Spatial Dimensions of Public Policy*. Oxford: Pergamon, pp. 86–103.

Howard, E. (1898) *Tomorrow: A Peaceful Path to Real Reform*. London: Swann Sonenschein (Reissued as *Garden Cities of Tomorrow*, 1902).

Howkins, A. (1991) *Reshaping Rural England: A Social History 1850–1925*. London: Harper Collins.

Hubbard, L. (1982) New Jobs in West Durham. Agricultural Adjustment Unit, University of Newcastle-upon-Tyne, Newcastle.

Huxley Report (1947) *Conservation of Nature in England and Wales*. Special Committee on Wildlife Conservation, Cmd 7122. London: HMSO.

Jack Committee (1961) *Rural Bus Services*. London: HMSO.

Jackson, A. (1973) *Semi-Detached London: Suburban Development, Life and Transport, 1900–1939*. London: George Allen and Unwin.

Jackson, V.J. (1968) *Population in the Countryside: Growth and Stagnation in the Cotswolds*. London: Frank Cass.

Jacobs, C.A. (1972) *Second Homes in Denbighshire*. Ruthin: Denbighshire County Council.

Jenkins, I. (1953) *The History of the Women's Institute Movement of England and Wales*. Oxford: Oxford University Press.

Joad, C.E.M. (1946) *The Untutored Townsman's Invasion of the Country*. London: Faber and Faber.

Johnston, R.J. (1966) Components of rural population change. *Town Planning Review*, **36** (4), pp. 279–293.

Johnstone, W.D., Nicholson, C., Stone, M. and Taylor, R.E. (1990) *Countrywork: A New Review of Rural Economic, Training and Employment*. Cirencester: ACRE/Planning Exchange.

Jones, G.E. (1984) *Modern Wales: A Concise History c 1485–1979*. Cambridge: Cambridge University Press.

JURUE (1981) The Social Effects of Rural Primary School Reorganization. JURUE, University of Aston.

Keith-Lucas, B. and Richards, P.G. (1978) *A History of Local Government in the Twentieth Century*. London: George Allen and Unwin.

Keynes, J.M. (1919) *The Economic Consequences of the Peace*. London: Macmillan.

Keynes, J.M. (1936) *The General Theory of Employment, Interest and Money*. London: Macmillan.

King, A.D. (1984) *The Bungalow: The Production of a Global Culture*. London: Routledge and Kegan Paul.

Law, C.M. (1967) The growth of urban population in England and Wales 1801–1911. *Transactions of the Institute of British Geographers*, **XLI**, p. 132.

Lawton, R. (ed.) (1978) *The Census and Social Structure: An Interpretive Guide to Nineteenth Century Censuses for England and Wales*. London: Frank Cass.

Lawton, R. (1983) Urbanization and population change in nineteenth century England, in Patten, J. (ed.) *The Expanding City*. London, Academic Press, pp. 179–224.

Lewis, G.J. (1991) Rural development in practice – the British experience, in Brunet, P. (ed.) *Regional Rural Development in Europe*. Caen: University of Caen, pp. 77–84.

Liberal Land Committee (1925) *The Land and the Nation*. London: Hodder and Stoughton.

Lievesley, K. and Maynard, W. (1992) *1991 Survey of Rural Services*. London: Rural Development Commission.

Little, J. (1987) Gender relations in rural areas: the importance of women's domestic role. *Journal of Rural Studies*, **3**, pp. 335–342.

Local Government Board (1919) *Manual on the Preparation of State-Aided Housing Schemes*. London: HMSO.

Lowe, P. (1989) The rural idyll defended: from preservation to conservation, in Mingay, G.E. (ed.) *The Rural Idyll*. London: Routledge, pp. 113–131.

Lowe, P., Bradley, T. and Wright, S. (1986) *Deprivation and Welfare in Rural Areas*. Norwich: GeoBooks.

Lowe, P., Cox, G., McEwen, M., O'Riordan, T. and Winter, M. (1986) *Countryside Conflicts: The Politics of Farming, Forestry and Conservation*. Aldershot: Gower.

Lowe, P. and Goyder, J. (1983) *Environmental Groups in Politics*. London: Allen and Unwin.

McAllister, G. and E.G. (eds.) (1945) *Homes, Towns and Countryside: A Practical Plan for Britain*. London: Batsford.

McCann, N.F. (1976) The background to the National Agricultural Advisory Service. *Journal of the Royal Agricultural Society of England*, **137**, pp. 51–59.

MacEwen, A. and M. (1982) *National Parks: Conservation or Cosmetics*. London: George Allen and Unwin.

McLaughlin, B. (1986) The rhetoric and reality of rural deprivation. *Journal of Rural Studies*, **2**, pp. 292–307.

McLaughlin, B. (1987) Rural policy into the 1990s: self-help or self-deception. *Journal of Rural Studies*, **3**, pp. 81–90.

Macnicol, J. (1986) The evacuation of school children, in Smith, H. (ed.) *War and Social Change: British Society in the Second World War*. Manchester: Manchester University Press, pp. 3–31.

Marsden, T., Murdoch, J., Lowe, P., Munton, R. and Flynn, A. (1993) *Constructing the Countryside*. London: UCL Press.

Marsh, J. (1982) *Back to the Land: The Pastoral Impulse in Victorian England from 1880 to 1914*. London: Quartet Books.

Marwick, A. (1976) *The Home Front: The British and the Second World War*. London: Thames and Hudson.

Massingham, H.J. (1945) Introduction, to Thompson, Flora *Lark Rise to Candleford*. Oxford: Oxford University Press, pp. 7–15.

Mawson, T.H. (1927) *The Life and Work of an English Landscape Architect*. London: Batsford.

Meadows, D.H., Meadows, D.L., Randers, J. and Behrens, W.W. (1972) *Limits to Growth*. London: Earth Island.

Miller, M. (1992) *Raymond Unwin: Garden Cities and Town Planning*. Leicester: Leicester University Press.

Minay, C. (1985) The Development Commission's Rural Industrial Development Programme: A Review of Progress 1945–1985. Department of Town Planning, Oxford Polytechnic, Oxford.

Minay, C. (1990) The Development Commission and English rural development, in Buller, H. and Wright, S. (eds.) *Rural Development: Problems and Practices*. Aldershot: Avebury, pp. 211–225.

Mingay, G.E. (ed.) (1981) *The Victorian Countryside*. 2 vols. London: Routledge and Kegan Paul.

Mingay, G.E. (ed.) (1989) *The Rural Idyll*. London: Routledge.

Mingay, G.E. (1990) The rural slum, in Martin, S. (ed.) *Slums*, Leicester: Leicester University Press, pp. 92–143.

Ministry of Agriculture, Fisheries and Food (1968) *A Century of Agricultural Statistics – Great Britain 1866–1966*. London: HMSO.

Ministry of Agriculture, Fisheries and Food (1972) *Forestry Policy*. London: HMSO.

Ministry of Agriculture, Fisheries and Food (1979) *Farming and the Nation*. Cmd 7458. London: HMSO.

Ministry of Food (1946) *How Britain was Fed in Wartime: Food Control, 1939–1945*. London: HMSO.

Ministry of Health (1919) *Manual on the Preparation of State-Aided Housing Schemes*. London: HMSO.

Ministry of Health (1944) *Rural Housing: Third Report of the Rural Housing Subcommittee of the Central Housing Advisory Committee*. London: HMSO.

Ministry of Housing and Local Government (1953) *Design in Town and Village*. London: HMSO.

Ministry of Housing and Local Government (1966) *Leisure in the Countryside, England and Wales*. Cmd 2928. London: HMSO.

Ministry of Housing and Local Government (1967) *Settlement in the Countryside: A Planning Method*. Planning Bulletin 8. London: HMSO.

Ministry of Information (1945) *Land at War: The Official History of British Farming, 1939–1944*. London: HMSO.

Ministry of Land and Natural Resources (1966) *Leisure in the Countryside, England and Wales*. Cmd 2928. London: HMSO.

Ministry of Town and Country Planning (1944) *The Control of Land Use*. Cmd 6537. London: HMSO.

Ministry of Town and Country Planning (1947) *Footpaths and Access to the Countryside: Report of the Special Committee (England and Wales)*. Cmd 7202. London: HMSO.

Minns, R. (1980) *Bombers and Mash: The Domestic Front, 1939–1945*. London: Virago.

Mitchell, G.D. (1950) Depopulation and rural social structure. *Sociological Review*, **42**, pp. 69–85.

Moore, D.C. (1981) The landed aristocracy, in Mingay, G.E. (ed.) *The Victorian Countryside*, London: Routledge and Kegan Paul, pp. 367–382.

Moore, N.W. (1969) Experience with pesticides and the theory of conservation. *Biological Conservation*, **1**, pp. 201–207.

Moreau, R.E. (1968) *The Departed Village: Berrick Salome at the Turn of the Century*. Oxford: Oxford University Press.

Morris, W. (1890) *News from Nowhere*. London: Longmans, Green.

Moseley, M.J. (1973) The impact of growth centres in rural regions. *Regional Studies*, **7**, pp. 57–94.

Moseley, M.J. (1974) *Growth Centres in Spatial Planning*. Oxford: Pergamon.

Moseley, M.J. (1979) *Accessibility: The Rural Challenge*. London: Methuen.

Moss, G. (1978) The village: a matter of life or death. *The Architects' Journal*, 18 January, pp. 100–139.

Moss, G. (1981) *Britain's Wasting Acres; Land Use in a Changing Society*. London: Architectural Press.
Murray, K. (1955) *Agriculture: History of the Second World War*, Vol. 6. London: HMSO.
Nairn, I. (ed.) (1955) Outrage. Special Number, *Architectural Review*, **117** (702), pp. 363–460.
National Park Committee (1931) *Report*. Cmd 3851. London: HMSO.
Nature Conservancy Council (1977) *Agriculture and Nature Conservation*. London: HMSO.
Nature Conservancy Council (1984) *Nature Conservation in Great Britain*. Peterborough: NCC.
Nature Conservancy Council (1986) *Nature Conservation and Afforestation in Britain*. Peterborough: NCC.
Newbold, H.B. (1942) *Industry and Rural Life*. London: Faber and Faber.
Newby, H. (1977) *The Deferential Worker: A Study of Farm Workers in East Anglia*. London: Allen Lane.
Newby, H. (1979/1985) *Green and Pleasant Land? Social Change in Rural England*. London: Hutchinson. Republished 1985 Wildwood House, London.
Newby, H. (1987) *Country Life: A Social History of Rural England*. London: Weidenfeld and Nicholson.
Newby, H., Bell, C., Rose, D. and Saunders, P. (1978) *Property, Paternalism and Power: Class and Control in Rural England*. London: Hutchinson.
New Townsmen (1918) *New Towns after the War*. London: Dent.
Nicholson, M. (1970) *The Environmental Revolution: A Guide for the New Masters of the World*. London: Hodder and Stoughton.
North, D. and Smallbone, D. (1993) Small Businesses in Rural Areas. Strategy Review Topic Paper 2, Rural Development Commission, London.
Northfield Committee (1979) *Report of the Committee of Inquiry into the Acquisition and Occupancy of Agricultural Land*. Cmd 7599. London: HMSO.
Offer, A. (1981) *Property and Politics 1870–1914*. Cambridge: Cambridge University Press.
Oliver, P., Davis, I. and Bentley, I. (1981) *Dunroamin: The Suburban Semi and Its Enemies*. London: Barrie and Jenkins.
Orwin, C.S. (1942) *Speed the Plough*. Harmondsworth: Penguin.
Orwin, C.S. (1945) *Problems of the Countryside*. Cambridge: Cambridge University Press.
Oxford Agricultural Economics Research Institute (1944) *Country Planning: A Study of Rural Problems*. Oxford: Oxford University Press.
Pahl, R.E. (1965) Urbs in Rure: The Metropolitan Fringe in Hertfordshire. Geographical Papers 2, London School of Economics, London.
Pahl, R.E. (1970) The social objectives of village planning, in Pahl, R.E. (ed.) *Whose City? and Further Essays on Urban Society*. Harmondsworth: Penguin, pp. 40–53.
Parker, D.J. and Penning-Rowsell, E.C. (1981) *Water Planning in Britain*. London: Allen and Unwin.
Parry, M.L., Bruce, A. and Harkness, C.E. (1982) Changes in the Extent of Moorland in the North York Moors. Department of Geography, University of Birmingham, Birmingham.
Patmore, J.A. (1970) *Land and Leisure in England and Wales*. Newton Abbot: David and Charles.

Patmore, J.A. (1983) *Recreation and Resources: Leisure Patterns and Leisure Places*. Oxford: Blackwell.
Pearson, L.F. (1991) *The People's Palaces: The Story of the Seaside Pleasure Buildings of 1870–1914*. Buckingham: Barracuda Books.
Pedley, W.H. (1942) *Labour on the Land: A Study of the Developments between the Two Great Wars*. London: King and Staples.
Pepper, D. (1984) *The Roots of Modern Environmentalism*. London: Routledge.
Perkin, H. (1989) *The Rise of Professional Society: England since 1880*. London: Routledge.
Perry, J. (1974) *British Farming in the Great Depression 1870–1914: A Historical Geography*. Newton Abbot: David and Charles.
Phillips, D.R. and Williams, A. (1983) *Rural Britain: A Social Geography*. Oxford: Blackwell.
Plowden Report (1967) *Children and Their Primary Schools*. Central Advisory Council for Education. London: HMSO.
Poole, K.P. and Keith-Lucas, B. (1994) *Parish Government 1894–1994*. London: National Association of Local Councils.
Porchester Lord (1977) *A Study of Exmoor*. London: HMSO.
Potter, C.A. (1986) Processes of countryside change in lowland England. *Journal of Rural Studies*, **2**, pp. 187–195.
Potter, C.A. (1993) Pieces in a jigsaw: a critique of the new agrienvironment measures. *ECOS*, **14** (1), pp. 52–54.
Potter, C.A., Burnham, C.P., Edwards, A., Gasson, R. and Green, B.H. (1991) *The Diversion of Land: Conservation in a Period of Farming Contraction*. London: Routledge.
Prothero, R.E. (1912) *English Farming Past and Present*. London: Longmans, Green.
Ravenstein, E. (1885) The laws of migration. *Journal of the Royal Statistical Society*, **48**, pp. 167–227.
Richards, S. (1994) *Wye College and Its World – A Centenary History*. Ashford: Wye College Press.
Robinson, G.M. (1990) *Conflict and Change in the Countryside*. London: Belhaven Press.
Rogers, A.W. (1976) Rural housing, in Cherry, G.E. (ed.) *Rural Planning Problems*. London: Leonard Hill, pp. 85–122.
Rogers, A.W. (1985) Local claims on rural housing: a review. *Town Planning Review*, **56**, pp. 36–80.
Rogers, A.W. (1987) Voluntarism, self-help and rural community development: some current approaches. *Journal of Rural Studies*, **3**, pp. 353–360.
Rogers, A.W. (1989) People in the countryside, in Mingay, G.E. (ed.) *The Rural Idyll*. London: Routledge, pp. 103–112.
Rogers, A.W. (1992) Legislating for the tied cottage: a study of the Rent (Agriculture) Act 1976. *Journal of the Royal Agricultural Society of England*, **153**, pp. 81–88.
Rogers, A.W. (1993) English Rural Communities: An Assessment and Prospect for the 1990s. Strategy Review Topic Paper 3, Rural Development Commission, London.
Rothman, B. (1982) *The 1932 Kinder Scout Trespass*. Timperley, Altringham: Willow Publishing.
Rural Development Commission (1993) *Rural Development Areas 1994*. London: Rural Development Commission.

Sackville-West, V. (1944) *The Women's Land Army*. London: Michael Joseph.
Sandbach, F.R. (1978) The early campaign for a National Park in the Lake District. *Transactions of the Institute of British Geographers*, New Series, **3** (4), pp. 498–514.
Sandford Lord (1974) *Report of the National Park Policy Review Committee*. London: HMSO.
Saville, J. (1957) *Rural Depopulation in England and Wales, 1851–1951*. London: Routledge.
Schumacher, F. (1973) *Small is Beautiful: Economics As If People Really Mattered*. London: Abacus.
Scott Report (1942) *Report of the Committee on Land Utilization in Rural Areas*. Cmd 6378. London: HMSO.
Self, P. and Storing, H. (1962) *The State and the Farmer*. London: Allen and Unwin.
Seymour, W.A. (ed.) (1980) *A History of the Ordnance Survey*. Folkestone: Dawson.
Sharp, T. (1932) *Town and Countryside: Some Aspects of Urban and Rural Development*. Oxford: Oxford University Press.
Sharp, T. (1936) *English Panorama*. London: Architectural Press.
Sharp, T. (1946) *The Anatomy of the Village*. Harmondsworth: Penguin.
Shaw, G.B. (ed.) (1889) *Fabian Essays in Socialism*. London: Fabian Society.
Shaw, J.M. (1976) Can we afford villages? *Built Environment*, **2**, pp. 135–137.
Shaw, J.M. (ed.) (1979) *Rural Deprivation and Planning*. Norwich: GeoBooks.
Sheail, J. (1975) The concept of National Parks in Great Britain 1900–1950. *Transactions of the Institute of British Geographers*, **66**, pp. 41–56.
Sheail, J. (1976*a*) *Nature in Trust; The History of Nature Conservation in Britain*. London: Blackie.
Sheail, J. (1976*b*) Coasts and planning in Great Britain before 1950. *The Geographical Journal*, **142** (2), pp. 257–273.
Sheail, J. (1981) *Rural Conservation in Inter-War Britain*. Oxford: Oxford University Press.
Sheail, J. (1983) Deserts of the moon: the Mineral Workings Act and the restoration of ironstone workings in Northamptonshire, England, 1936–1951. *Town Planning Review*, **54** (4), pp. 405–424.
Sheail, J. (1991) *Power in Trust: The Environmental History of the Central Electricity Generating Board*. Oxford: Clarendon Press.
Sheail, J. (1992*a*) The South Downs and Brighton's water supplies: an interwar study of resource management. *Southern History*, **14**, pp. 93–111.
Sheail, J. (1992*b*) The 'amenity' clause: an insight into half a century of environmental protection in the United Kingdom. *Transactions of the Institute of British Geographers*, **17**, pp. 152–165.
Sheail, J. (1993*a*) Sewering the English suburbs: an interwar perspective. *Journal of Historical Geography*, **19** (2), pp. 433–447.
Sheail, J. (1993*b*) Green history: the evolving agenda. *Rural History*, **4** (2), pp. 209–223.
Sheail, J. (1993*c*) The agricultural pollution of watercourses: the precedents set by the beet-sugar and milk industries. *Agricultural History Review*, **41** (1), pp. 31–44.
Sheail, J. (1995) Dower in context: National Parks and Town and Country Planning in Britain. *Planning Perspectives*, **10** (1), pp. 1–6.

Sheppard, J.A. (1962) Rural population changes since 1851: three sample studies. *Sociological Review* (NS), **10**, pp. 81–95.
Shoard, M. (1980) *The Theft of the Countryside*. London: Temple Smith.
Shoard, M. (1987) *This Land is Our Land*. London: Grafton.
Shucksmith, M. (1981) *No Homes for Locals?* Aldershot: Gower.
Shucksmith, M. (1990) *Housebuilding in Britain's Countryside*. London: Routledge.
Smailes, A.E. (1944) The urban hierarchy in England and Wales. *Geography*, **29**, pp. 41–51.
Smart, G. and Anderson, M. (1990) *Planning and Management of Areas of Outstanding Natural Beauty*. Cheltenham: Countryside Commission.
Smith, H.L. (1986) *War and Social Change: British Society in the Second World War*. Manchester: Manchester University Press.
Stacey, M. (1960) *Tradition and Change: A Study of Banbury*. Oxford: Oxford University Press.
Stacey, M., Batston, E., Bell, C. and Murcott, A. (1975) *Power, Persistence and Change: A Second Study of Banbury*. London: Routledge and Kegan Paul.
Stamp, L.D. (1946) *The Land of Britain and How It Is Used*. London: Longmans, Green.
Stamp, L.D (1947) Wartime changes in British agriculture. *Geographical Journal*, **109**, pp. 39–57.
Stamp, L.D. (1962) *The Land of Britain: Its Use and Misuse*, 3rd ed. London: Longmans, Green.
Standing Conference of Rural Community Councils (1978) *The Loss of Rural Services*. London: National Council of Social Service.
Stansfield, K. (1981) Thomas Sharp 1901–1978, in Cherry, G.E. (ed.) *Pioneers in British Planning*. London: Architectural Press, pp. 150–176.
Stapledon, R.G. (1935) *The Land: Now and Tomorrow*. London: Faber and Faber.
Steers, J.A. (1946) *The Coastline of England and Wales*. Cambridge: Cambridge University Press.
Stephenson, T. (1989) *Forbidden Land: The Struggle for Access to Mountain and Moorland*. Manchester: Manchester University Press.
Stockford, D. (1978) Social services provision in rural Norfolk, in Moseley, M.J. (ed.) *Social Issues in Rural Norfolk*. Norwich: University of East Anglia, pp. 59–75.
Sykes, J.D. (1981) Agriculture and science, in Mingay, G.E. (ed.) *The Victorian Countryside*. London: Routledge and Kegan Paul, pp. 260–272.
Symes, D.G. (1981) Rural communities in Great Britain, in Durand-Drouhin, J.-L. and Szwengrub, L.-M. (eds.) *Rural Community Studies in Europe*, **1**. Oxford: Pergamon Press.
Thomas, D. (1970) *London's Green Belt*. London: Faber and Faber.
Thomas, F.G. (1939) *The Changing Village: An Essay on Rural Reconstruction*. London: Nelson.
Thompson, F. (1945) *Lark Rise to Candleford*. Oxford: Oxford University Press.
Thompson, F.M.L. (ed.) (1982) *The Rise of Suburbia*. Leicester: Leicester University Press.
Thorburn, A. (1971) *Planning Villages*. London: Estates Gazette.
Titmuss, R. (1950) *Problems of Social Policy*. London: HMSO.
Titmuss, R. (1958) *Essays on 'The Welfare State'*. London: Allen and Unwin.
Townsend, A. (1991) New forms of employment in rural areas: a national

perspective, in Champion, T. and Watkins, C. (eds.) *People in the Countryside: Studies of Social Change in Rural Britain*. London: Paul Chapman, pp. 84–95.

Tracy, M. (1989) *Government and Agriculture in Western Europe 1880–1988*. Hemel Hempstead: Harvester Wheatsheaf.

Trevelyan, G.M. (1938) Foreword. *The Case for National Parks in Great Britain*. Standing Committee on National Parks. London: HMSO.

Tricker, M.J. (1983) Rural education services: the social effects of reorganization, in Clark, G., Groenendijk, J. and Thissen, F. (eds.) *The Changing Countryside*. Norwich: GeoBooks, pp. 111–119.

Tricker, M.J. and Mills, L. (1987) Education services, in Cloke, P.J. (ed.) *Rural Planning: Policy into Action?* London: Harper and Row, pp. 37–55.

Tudor Walters, Sir J. (1918) *Report of the Committee appointed by the President of the Local Government Board and the Secretary for Scotland to consider Questions of Building Construction in connection with the Provision of Dwellings for the Working Classes in England, Wales and Scotland*. Cmd 9191. London: HMSO.

Unwin, R. (1902) *Cottage Plans and Common Sense*. London: Fabian Society.

Unwin, R. (1912) *Nothing gained by Overcrowding: How the Garden City Type of Development may benefit both the Owner and the Occupier*. London: Garden Cities and Town Planning Association.

Unwin, R. and Parker, B. (1901) *The Art of Building a Home*. London: Longmans, Green.

Uthwatt Report (1942) *Expert Committee on Compensation and Betterment*. Final Report, Ministry of Works and Planning. Cmd 6386. London: HMSO.

Waller, R. (1962) *Prophet of the New Age: the Life and Thought of Sir George Stapledon*. London: Faber.

Walton, J.K. (1983) *The English Seaside Resort: A Social History 1750–1914*. Leicester: Leicester University Press.

Wannop, U. and Cherry, G.E. (1994) The development of regional planning in the United Kingdom. *Planning Perspectives*, **9**, pp. 29–60.

Ward, B. and Dubos, R. (1972) *Only One Earth: The Care and Maintenance of a Small Planet*. Harmondsworth: Penguin.

Ward, C. (1988) *The Child in the Country*. London: Robert Hale.

Ward, C. and Hardy, D. (1986) *Goodnight Campers! The History of the British Holiday Camp*. London: Mansell.

Ward, S. *et al.* (1982) *Seasons of Change: Rural Life in Victorian and Edwardian England*. London: George Allen and Unwin.

Wells, H.G. (1902) *Anticipations*. London: Chapman and Hall.

Westmacott, R. and Worthington, T. (1974) New Agricultural Landscapes. CCP 76, Countryside Commission, Cheltenham.

Whatmore, S. (1991) *Farming Women: Gender, Work and Family Enterprise*. London: Macmillan.

Whetham, E.H. (1952) *British Farming, 1939–1949*. London: Nelson.

Wibberley, G.P. (1959) *Agriculture and Urban Growth: A Study of the Competition for Rural Land*. London: Michael Joseph.

Wibberley, G.P. (1982) *Countryside Planning: A Personal Evaluation*. Ashford: Wye College, University of London.

Wibberley, G.P. (1985) The famous Scott Report: a text for all time? *The Planner*, **7** (4), pp. 13–20.

Wilkinson, P. (1992) *Thames Chase Plan – Draft for Consultation*. London: Thames Chase.

Williams-Ellis, C. (1928) *England and the Octopus*. London: Bles.
Williams, R. (1973) *The Country and the City*. London: Chatto and Windus (St Albans: Granada, 1975).
Williams, W.H. (1965) *The Commons, Open Spaces and Footpaths Preservation Society 1865–1965: A Short History*. London: The Society.
Williams, W.M. (1963) *A West Country Village: Ashworthy*. London: Routledge and Kegan Paul.
Winegarten, A. and Acland-Hood, M. (1978) British agriculture and the 1947 Agriculture Act. *Journal of the Royal Agricultural Society of England*, **139**, pp. 74–82.
Woodruffe, B.J. (1976) *Rural Settlement Policies and Plans*. Oxford: Oxford University Press.
Wright, N.C. (1941) The milk supply in time of war. *Journal of the Royal Agricultural Society of England*, **101** (11), pp. 44–56.
Wright, S. (1990) Development theory and community development practice, in Buller, H. and Wright, S. (eds.) *Rural Development: Problems and Practices*. Aldershot: Avebury, pp. 41–63.
Wright, S. (1992) Image and analysis: new directions in community studies, in Short, B. (ed.) *The English Rural Community: Image and Analysis*. Cambridge: Cambridge University Press, pp. 195–217.
Young, M. (1982) *The Elmhirsts of Dartington – The Creation of a Utopian Community*. London: Routledge and Kegan Paul.
Young, K. (ed.) (1989) *New Directions for County Government*. London: Association of County Councils/Institute of Local Government Studies, University of Birmingham.

INDEX

For Product Safety Concerns and Information please contact our EU
representative GPSR@taylorandfrancis.com
Taylor & Francis Verlag GmbH, Kaufingerstraße 24, 80331 München, Germany

www.ingramcontent.com/pod-product-compliance
Ingram Content Group UK Ltd.
Pitfield, Milton Keynes, MK11 3LW, UK
UKHW020158200926
475702UK00004B/517
9781138344204